GW00778096

NOTES AND COMMENTS ON THE COMPOSITION OF TERRESTRIAL AND CELESTIAL MAPS

By JOHANN HEINRICH LAMBERT

From Part III of
Contributions to the Practice of Mathematics and its Application, Berlin, 1772

Translated and Introduced by
Waldo R. Tobler
Geography Department
University of California, Santa Barbara

This monograph first appeared as
University of Michigan Department of Geography publication No. 8,
edited by Professors T. Detwyler and J. Kolars, in Ann Arbor in 1972
on the 200th anniversary of its original publication.

Originally published by
University of Michigan
Ann Arbor, Michigan, USA
uwpress.wisc.edu

Esri Press, 380 New York Street, Redlands, California 92373-8100

 First edition 2011

15 14 13 12 11   1 2 3 4 5 6 7 8 9 10

Printed in the United States of America

Library of Congress Cataloging-in-Publication Data

Lambert, Johann Heinrich, 1728-1777.
Notes and comments on the composition of terrestrial and celestial maps / by Johann Heinrich Lambert.
p. cm.
Includes bibliographical references and index.
ISBN 978-1-58948-281-4 (pbk. : alk. paper)
1. Map projection. I. Title.
GA115.L2213 2011
526'.8--dc22   2010051306

Ask for Esri Press titles at your local bookstore or order by calling 800-447-9778, or shop online at www.esri.com/esripress. Outside the United States, contact your local Esri distributor or shop online at www.eurospanbookstore.com/Esri.

Esri Press titles are distributed to the trade by the following:

In North America:

Ingram Publisher Services
Toll-free telephone: 800-648-3104
Toll-free fax: 800-838-1149
E-mail:
customerservice@ingrampublisherservices.com

In the United Kingdom, Europe, Middle East and Africa, Asia, and Australia:

Eurospan Group
3 Henrietta Street
London WC2E 8LU
United Kingdom

Telephone: 44(0) 1767 604972
Fax: 44(0) 1767 601640
E-mail: eurospan@turpin-distribution.com

ABOUT THIS REPRINTING

Esri Press is honored to republish Waldo R. Tobler's translation of Johann Heinrich Lambert's seminal writings on map projections—*Notes and Comments on the Composition of Terrestrial and Celestial Maps*. Tobler first published his translation in 1972 on the 200th anniversary of Lambert's original publication (in German). This new edition updates the cover design and includes a photograph by the author. Like its predecessor, this edition contains a biography and bibliography by Hans Maurer (written in 1931). Tobler has updated his own listing of selected references in the back of the book, and a short index has also been added.

The text has been edited for consistent style; minor wording and punctuation changes were made in some instances. Time references were left as they were originally written. Original bibliographic style has been preserved in most cases.

Esri Press is pleased to make this cartographic classic available to readers once again.

CONTENTS

PREFACE

The subject of map projections has seen remarkable contributions from many remarkable individuals. Johann Heinrich Lambert (1728–77) was one of the subject's foremost contributors. His work is generally considered to constitute the beginning of the modern period in mathematical cartography. An English rendition of his *Notes and Comments on the Composition of Terrestrial and Celestial Maps* on the 200th anniversary of its publication is therefore most appropriate.

The subject is an ancient one, going back at least two millennia. The hypothesis of a spherical heaven or earth, of course, was required before the problem was recognized. Difficulties in the construction and use of actual spheres most probably provided the motivation for the first map projection. Ancient celestial maps were constructed on the gnomonic or stereographic projections, while the concept of latitude and longitude is also more than twenty centuries old. A rectangular grid using this scheme is an obvious choice for indexing geographical data and this also is among the oldest of projections.

Knowledge of the next advances in map projection comes mainly from manuscript copies of Ptolemy's *Geographia* (circa A.D. 150). One of his chapter headings reads: "How the habitable earth can be shown on a plane so that its measurements are in keeping with its spherical shape." Ptolemy describes several specific projections, but the fundamental principle he presents is more important: Choose a projection so as to reduce incorrect distances. A recurring theme in Ptolemy's work, and one that continues to this day, is that a flat map should attempt to preserve some spherical property. Ptolemy also knew, or guessed, that all spherical distances cannot be correct on a flat map, and states that "the difficulty is evident."

For several centuries following Ptolemy there appeared maps that contain no indication of the latitude and longitude graticule. These include the medieval mappae mundi, the so-called T-in-O maps, and the early portolan charts. Later, after A.D. 1400, Ptolemy's writings were translated and gained high repute. The roundness of the earth was again recognized; latitude and longitude reappeared on maps, and several dozen new map projections were invented. With one exception, the new projections on an individual basis were not particularly influential, since innumerable map projections are possible. A few preserve important spherical properties and are valuable projections still in use today. They seem, however, to have been invented on a somewhat ad hoc basis, as particularistic schemes for drawing the spherical graticule. Undergraduate students invent new projections today by the same method.

The most important projection of the period was that devised by Gerardus Mercator (1569). This is still used by sailors to determine magnetic bearings of destinations too far away to be reached by line-of-sight sailing. In a modern interpretation, the curves that interested Mercator are logarithmic spirals on a sphere, which he succeeded in converting into straight lines on a map. His map projection is not to be used for indexing nor for the measurement of long distances (which are distorted); rather it is a nomograph to solve a particular problem in navigation graphically. Algebraically Mercator's projection transforms the problem of finding the intersection angle between a spherical spiral and a meridian into the much simpler problem of finding the plane intersection angle between two lines. The preservation of a spherical property on a flat map is again implied, as with Ptolemy, but in this case it is not distance that is invariant. Requiring that the exaggeration of lengths along the meridians be the same as the exaggeration along the parallels makes the projection conformal. Some thirty years after the publication of Mercator's map, E. Wright's book on the correction of navigational errors, *Certain Errors in Navigation Detected and Corrected*, more fully explained Mercator's method. But several decades went by before sailors were convinced of the value of the new charts. Now it is difficult for them to see much value in any other type of map.

In the sixteenth century one can find a description of methods for constructing maps in the English edition of Varenius' famous *General Geography*. From this, one learns that maps should show places in their proper position with respect to latitude and longitude, that the magnitudes of regions be in proper proportion to their values on the earth, and that all places be in the same situation to each other as on the earth. Varenius asserts that the first requirement can be met; that the second can be met only if the eye be conceived to be infinitely remote from the earth; and the third can be met only for maps covering small areas. Reading further in the text one finds an emphasis on instructions for several of the perspective azimuthal projections, but Mercator's projection is also discussed. The treatment of Ptolemy's projections is rather vague. A trapezoidal projection and the square projection are mentioned, but several other existing projections are neglected (e.g., the sinusoidal, and those of Werner). The preservation of surface areas is an implied desideratum. We must recall that Ptolemy cited preservation of distances as a criterion but not explicitly the preservation of area, the separability of these two concepts apparently not having been recognized in his time.

Moving on to the next century, one finds that an article by P. Murdock, "On the Best Form of Geographical Maps," in the *Philosophical Transactions of the Royal Society* for 1758, contains the question: "What is the construction of a particular map, that shall exhibit the superficial and linear measures in their truest proportions?" The answer is surprisingly good, for Murdock suggests what would now be known as secant conical projections.

In 1772 Johann Heinrich Lambert published his treatise "Anmerkungen and Zusätze zur Entwerfung der Land and Himmelscharten." This work is outstanding for several reasons. Lambert specifically stated the objective of preservation of spherical properties as a desideratum for plane maps. He continued by listing some of the possible properties to be preserved including the preservation of local similarities. Although Mercator's projection, and the stereographic projection, already did this, Lambert treated the subject much more generally. An even greater innovation on Lambert's part was the attempt to characterize these properties in analytical form (as nonlinear partial differential equations) and to suggest general solutions, including oblique cases. He, thus, literally invented an infinite number of specific map projections. Lambert's analytical approach has characterized all important work on map projections since that date. Perhaps of equal significance is the fact that Lambert brought the subject to the attention of outstanding mathematicians. In quick succession there appeared treatises by Euler, Lagrange, and, somewhat later, Gauss. Euler, in addition to his projections, proved analytically the impossibility of preserving all spherical properties on a flat map. His proof consists basically in demonstrating that the sum of the interior angles of a triangle, whose sides are geodesic arcs, is 180 degrees for a plane triangle and is greater than this for a spherical triangle. If the sum is greater than 180 degrees on a plane then the sides cannot be geodesic, and the converse.

Gauss' contributions were fundamental. Characteristically he did not attack so simple a problem as representing the surface of a sphere on a plane, but rather considered the general question of representing one arbitrary surface on another arbitrary surface in such a manner that similitude relations are maintained. He thus gave a general solution using complex variables to the problem, posed by Lambert, of conformal representations. Gauss continued by creating the subject of differential geometry, now basic to advanced work on map projections, and applied the method of least squares to solve geodetic problems.

The remainder of the nineteenth century was a period of great mathematical activity, little of which is reflected in the geographical literature on the subject of map projections. Specific projections were invented and solutions were sought to such comparatively minor, though difficult, problems as finding projections that have specifically shaped meridians or parallels. At that time most of the major problems appeared to have been solved. In 1881, however, Tissot published his *Mémoire* on projections in which he organized the study of the distortion on map projections. In relatively simple terms, he asked: If the objective is to preserve some spherical property on a flat map, what happens to the other properties when this is done? How are the angles distorted on equal-area projections, and so on? Tissot's indicatrix allows such questions to be answered, in a local sense, with comparative ease. Given his simple method of determining distortion, the objective could thereafter be modified somewhat. The newer problem might be stated: in addition to preservation of a spherical property (or properties) choose a projection that also minimizes the distortion of other properties. In other words, of all possible conformal projections, the most appropriate one is that which minimizes areal exaggeration for the region in question, or find a projection that balances areal and angular distortion. Several varieties of solutions are now available.

The majority of the developments in the twentieth century have continued in the same vein. The search for least-distorted projections continues, and more complicated oblique projections of the ellipsoid are being applied to topographical maps. The general value of conformal projections for numerous types of problems has been recognized. Plane coordinate systems, defined by a corresponding map projection, have proven their utility for local surveying purposes and for regional, economic, and demographic studies. Projections which do not preserve any specific spherical property, but which attempt to minimize distortion of several properties, are increasingly popular for atlases and wall maps. Many unusual projections also have been investigated. These are generally based on preservation of specific, often obscure, features of the spherical geometry (retro-azimuthals, for example), or attempt to simplify (i.e., make linear) some spherical relation. A recent map of this type represents the ground trace of an orbiting satellite—which is an oscillating curve on Mercator's projection—as a straight line of length proportional to transit times. This feat is performed by bending the meridians into S-shaped curves. Aerial photographs, radar, and earth satellite imagery all yield representations which can be treated within the classical geometrical framework of map projections. Recent exploration of celestial space has created a need for new maps, including maps of the moon. One projection chosen for this purpose was invented in the 1860s, but had heretofore not been employed for any actual maps. The theoretical advantage of this projection was not recognized as applying to any practical problems until 100 years after its invention. Digital computers have greatly simplified the calculation and drawing of maps, but not, as yet, the choice of map projection. It also has recently been shown that the perspective projections (rarely used for actual maps) have great advantages for digital processing and storage of large inventories of geographic information.

Within a broader framework, the subject of map projections is rather inextricably related to the general intellectual history of the western world. The notion of a spherical (later ellipsoidal) earth, and Mercator's contribution to the exploration of this planet are perhaps the best known examples. As a branch of projective geometry, map drawing interested artists of the caliber of Albrecht Dürer and Leonardo da Vinci. The philosopher Roger Bacon designed a map projection. Mathematics owes a great debt to geography because the idea of a mapping—a correspondence between entities—is now fundamental to almost all mathematics. The several branches of physics make frequent use of conformal mappings. The biologist D'Arcy Thompson has used transformations in an imaginative way to illuminate relations between species.

Of course not all of this is due to Lambert, but his was the insight that recognized that more lay hidden in the subject than had before been perceived. Though his successors clearly went beyond him, consider what successors they were! Euler, Lagrange, Gauss, even Riemann might be included. Euler's paper on projections is dated 1777; that of Lagrange 1779; Gauss' can be assigned to 1825. Neither Euler nor Gauss refers to Lambert's paper, but then they hardly ever refer to anybody in their works. Euler surely must have known of Lambert's work for they were together at the Academy in Berlin for several years. Lambert specifically mentions his connection to Lagrange. Both Albers and Mollweide refer to Lambert's work in their own 1805 papers on map projections in Zach's *Monatiche Correspondenz.* Gauss contributed several papers to this journal, and in general displayed a great interest in topics, especially astronomical, which also concerned Lambert. It is difficult to imagine that Gauss was unfamiliar with Lambert's work on projections, though I am not aware of hard evidence for a direct connection. Euler's projections are well known in Russia and are frequently used there, although they are not well known in the English speaking world. Lagrange's projection (actually a variant of a projection by Lambert) is known to most mathematicians but hardly ever used for actual maps. Gauss' work on projections is generally used for the geodetic calculation of triangulations and less often for actual maps. Thus, the practical cartographer works with one of Lambert's projections.

Five distinct projections can definitely be attributed to Lambert. These are the cylindrical equal-area projection, the azimuthal equal-area projection, the conical equal-area projection with one standard parallel, the transverse Mercator projection, and the conformal conical projection with two standard parallels. Of these, the last mentioned forms the basis for the official plane coordinate system in many states. It thus has a legal status for surveying and for the description of the location of real property. Naturally, it is the best known of the several Lambert Projections, but was not in wide use until sometime after 1920, when its advantages came to be recognized. The transverse Mercator projection, increasingly important for purposes similar to those of Lambert's conformal conic, is not usually attributed to Lambert but rather to Gauss and to later geodesists who provided more practical formulae for calculations involving the ellipsoid. Perhaps it required Gauss' genius to recognize the value of this projection of Lambert's.

Lambert's equal-area conical projection has been rendered more or less obsolete by that of Albers', which however differs from that of Lambert only by specification of slightly different boundary conditions in order to yield two standard parallels. The azimuthal equal-area projection of Lambert is used frequently in atlases and appears to be one of the most popular for this purpose. For hemispheric maps it does not have great angular distortions. The cylindrical equal-area projection discovered by Lambert does not appear to be much used although it provides an equal-area rectangular grid of parallels and meridians for the entire world, and the angular distortion is relatively small.

Lambert's achievements and contributions in other fields seem to have been at least as great as in cartography. In philosophy he disagreed with Kant and felt that geometry should be based on experience, not insight. The similarity, which probably should not be pushed too far, to later work by Gauss, Riemann, and Einstein is amazing. To mathematicians he is chiefly famous for almost perceiving non-Euclidean geometry. "Almost," of course, is not good enough, but he remarked on the similarity to spherical geometry obtained by replacing the parallel postulate, and conjectured on the existence of the pseudosphere now associated with the geometry of Lobachevsky and Bolyai. This work of Lambert's is today considered among the important precursors to non-Euclidean geometry. He also contributed to the discovery of the irrationality of π, thus helping Lindemann to put the circle squarers out of business, and he contributed to the subject of hyperbolic functions. Several theorems bear his name. His works on projective geometry are considered to have been unsurpassed, though little known, for many years. In logic he advocated a logical calculus, presumably like that later developed by Boole, and his numbering of concepts is hauntingly reminiscent of a notion used by Gödel in his famous proof. Lambert's achievements do not end there, as can be seen from the biography and bibliography provided in 1931 by H. Maurer, and appended here with the help of the American Geographical Society of New York and by permission of the International Hydrographic Bureau. A concise description of each reference's subject matter has been appended in brackets for the reader's convenience. A more recent publication concerning Lambert is by F. Lowenhaupt, ed., *J. H. Lambert, Leistung und Leben*, 110 pp., Mülhausen, 1943. The *Allgemeine Deutsche Biographie* may also be consulted. For the reader who wishes to pursue the subject of map projections further I have prepared a short list of references that has been placed at the end of the present volume.

Lambert's original work, "Anmerkungen und Zusätze zur Entwerfung der Land-und Himmelscharten," appeared as pages 105–199 of the third part of his *Beyträge zum Gebrauche der Matematik und deren Anwendung*, Verlag der Buchhandlung der Realschule, Berlin, 1772. An original copy of this work is available in the library of the University of Michigan, but my translation has been prepared mostly from A. Wangerin's rendition that appeared as Number 54, *Ostwald's Klassiker der exakten Wissenschaften*, Leipzig, 1894. I have used the copy donated to the University of Michigan by A. Ziwet. Wangerin's valuable commentary was used to interpret several passages in the text, and his rendition of the formulae and figures has been retained. The minor corrections over the original are essentially typographical or arithmetical and no other changes have been made. Lambert's clear and direct style made the translation an enjoyable task.

Waldo R. Tobler
Ann Arbor, Michigan
1972

Second printing (2011)

The creation of this version of my 1972 translation of Lambert's 1772 classic has allowed for the correction of several typographic errors. It has also allowed me to take advantage of reviews by R. Adler, H. Stoughton, and T. Wray and to insert a few changes and add some supplementary remarks and a short index of the projections described.

My further acknowledgements at this point include John Kolars and Tom Detwyler who published the original print edition as number eight in the Michigan Geographical Publications series. Jessica Edwards designed the initial cover and Judith DeBeaumont did most of the typing for that edition. Sarah Bow, under the direction of Jeff Hemphill, in the Geography Department at the University of California, Santa Barbara, scanned and did the preliminary edit of the current version. Professor Emeritus A. J. Kimerling of Oregon State University also suggested corrections to the manuscript. And I have added a few photographs from my July 1972 visit to Mulhouse.

One of the reviewers lamented that I did not give Lambert sufficient credit for his contributions to the subject of map projections. Specifically, he indicated that I should have stressed more strongly that Lambert was the first to consider in full generality the problem of conformal mapping of the sphere, and also that the transverse Mercator projection and that the projection known by Lagrange's name are due to Lambert. In the case of the transverse Mercator, Lambert must be recognized as the true inventor, but it was further developed by geodesists, including Gauss, and is now widely used in the version known throughout the world as the Universal Transverse Mercator (UTM) projection. In the Lagrangian case the criticism is perhaps justified. Rereading Lambert's work it is clear that he invented an entire family of projections, moving in a continuous fashion from the equatorial Mercator to the stereographic with the conformal conic and the projection attributed to Lagrange in between. Lambert himself however noted that Lagrange was a somewhat better mathematician, and this is evident from Lagrange's more fluid use of complex variables in his own monograph on map projections. Wangerin, in his commentary accompanying the republication of Lambert's classics paper suggests that Lambert did not quite know what to do with the equations for the general conformal mappings and that Lagrange went considerably beyond Lambert in this respect. And Gauss gave the general solution to conformal transformations in his paper of 1822.

In spite of the fact that sections I, II, and III of the text are a bit tedious, the reader should continue on since things improve in importance thereafter. Although a few parts of sections 74 through 83 involve many of Lambert's series expansions, the content gets better again as one continues reading. Both in the case of conformal projections and in the case of equivalent (equal area) projections Lambert was the first to clearly state, and importantly, write down, conditions in equation form, for both equality of area and of angles. He then developed several new projections of both types. In the case of equal area projections, he did this in ten short paragraphs. Although the equal-area conic with two standard parallels is attributed to Albers it is also clear that the procedure used by Lambert for the conformal conic projection in § 53 and § 54 can easily be applied to an equal area conic. Lambert's azimuthal equal area projection has now been chosen as the preferred basis for statistical maps for the European Union.

This same reviewer objected to my statement regarding Lambert's "helping Lindemann to put the circle squarers out of business," A recent (2004) mathematical history Internet page[1] points out that "Lambert was the first to provide a rigorous proof that π is irrational" and that he conjectured that "... e and π are transcendental. This was not proved for another century..." (also see Penrose 2005, pages 35, 43, 422). The same cited Internet page contains thirty-seven additional publications on Lambert's contributions, mostly in reference to his mathematical

1 http://www-history.mcs.st-andrews.ac.uk/history/Mathematicians/Lambert.html (accessed August 2010)

achievements. Of these, seventy-eight percent appeared after 1972, when the original version of my translation appeared. In relation to Gauss, Pesic in the introduction to the recent Dover edition to the "General Investigations of Curved Surfaces" in note 4 (page ix) suggests that "The young Gauss used Johann Lambert's table of prime numbers...." This is not a direct connection to Lambert's work on map projections, but it is not surprising that Gauss was cognizant of Lambert.

For readers who wish to read the (German language) Wangerin edition of Lambert's monograph, a scanned version of this is now available for downloading at the University of Michigan library Web site. Wangerin's appended remarks clarify several of Lambert's allusions to contemporary individuals and publications that I did not translate and comments on, and expansion, of some of the equations. The Internet also provides reference to a large number of descriptions of Lambert's work

The three-and-a-half decades since the 1972 edition have seen remarkable changes in the resources for the subject of map projections. Consequently, my list of references in this area is expanded slightly to incorporate a few items that have appeared in the intervening years; a recent search on Google resulted in 87,200 hits on the subject of "Map Projection." A majority of these seem to be elementary and conventional tutorials but some go into greater detail and are innovative. Most importantly, the computer revolution is quite strongly felt, and includes large quantities of geographic information recorded on computers in latitude-longitude coordinates along with equation algorithms, thus permitting virtually instantaneous creation of geographical maps on a variety of projections. Equally importantly, the Internet allows one access to such resources as John Snyder and Harry Steward's 1997 bibliography[2] with reference to some 2,995 publications in the field from circa 1500 to 1996. Assuming the 1980–90 production of approximately forty-eight new references per year, more than 825 papers would now need to be added to this bibliographic inventory. I have not attempted to do this.

In recent years recognition is made of the importance of a third additional class of map projection, beyond conformality and equivalence, namely that of *density preserving* projections, a generalization of the mathematics of equal area projections. This has been most notably captured in the cartograms of the "Atlas of the Real World." Important advances have also been in the area of *optimal projections*, taking

2 U.S. Geological Survey Bulletin 1856

advantage of Chebshev's result combined with digital computers. This period has not yet ended so that the coming years can be expected to see further advances in this direction. Such use is when one wishes to render, in great fidelity, the classical geometric properties. But it has also been recognized that map projections can be used in a manner similar to the use of logarithmic (or other variants of) graph paper to simplify, or render linear, spatial relations. Thus one can think of different map projections as different ways of producing different types of graph paper to be used as nomographs for assistance in solving different spatial problems, and not just metrically accurate depictions of the earth's surface. This mathematical paradigm of "Transform-Solve-Invert" (Eves 1980) is how Mercator used his famous anamorphose, but this also can be used more widely. It is used, for example, in aerodynamics in conjunction with mappings of wing shapes, and in several other fields.

As an aside, in addition to the conventional list of properties to be preserved, it is worth separate mention that the preservation of the two dimensionality of the earth's surface is required. Curiously, many authors still incorrectly refer to the map projection problem as one of going from a 3D spherical object to a plane. The problem is one of going from the 2D surface of a curved object to a 2D plane, and all map projections preserve this spherical, or spheroidal, property.

Finally, I am grateful for the leadership of Jack and Laura Dangermond, in devoting effort to the preservation of cartographic heritage by publishing documents of historical importance. I am also indebted to Esri Press under Peter Adams and his fine staff, including Monica McGregor, Julia Nelson, and David Boyles for their dedication and elegant production of a manuscript of more than usual difficulty.

Waldo Tobler
Santa Barbara
December 2010

The first page of the original text

Anmerkungen und Zusätze zur Entwerfung der Land- und Himmelscharten.

§. 1.

Man giebt überhaupt mehrere Bedingungen an, denen eine vollkommene Landcharte Genüge leisten soll. Sie soll 1. die Figur der Länder nicht verunstalten. 2. Die Größe der Länder sollen auf der Charte ihre wahre Verhältniß unter sich behalten. 3. Die Entfernungen jeder Oerter von jeden andern sollen ebenfalls in Verhältniß der wahren Entfernungen seyn. 4. Was auf der Erdfläche in gerader Linie, das will eigentlich sagen auf einem größten Circul der Sphäre, liegt, das soll auch in der Landcharte in gerader Linie liegen. 5. Die geographische Länge und Breite der Oerter soll auf der Charte leicht können gefunden werden rc. Das will nun überhaupt sagen, die Landcharten sollen in Absicht auf ganze Länder, ganze Welttheile oder die ganze Erdfläche durchaus eben das seyn, was ein Grundriß in Absicht auf ein Haus, Hof, Garten, Feld, Forst rc. ist. Dieses würde nun ganz wohl angehen, wenn die Erdfläche eine ebene Fläche wäre. Sie ist aber eine Kugelfläche, und damit läßt sich nicht allen

NOTES AND COMMENTS ON THE COMPOSITION OF TERRESTRIAL AND CELESTIAL MAPS

By JOHANN HEINRICH LAMBERT

From Part III of
Contributions to the Practice of Mathematics and its Application, Berlin, 1772

TABLE OF CONTENTS

INTRODUCTION

§ 1

Several conditions which an adequate map should satisfy have been recognized. It should: (1) not disturb the shape of countries; (2) countries should maintain their true relative sizes; (3) the distance of every place from every other place should also be proportional to the true distance; (4) places lying on a straight line on the earth's surface, that is, on a spherical great circle, should also lie on a straight line on the map; (5) the geographical latitude and longitude of places should be easily found on the map; and so on. In summary, a map should bear the same relation to countries, to hemispheres, or even to the entire earth, as do engineering drawings to a house, yard, garden, field, or forest. This would work if the surface of the earth were a flat surface. But it is a spherical surface, and all of the requirements cannot be satisfied simultaneously, and it is therefore necessary to emphasize one or several especially valuable requirements at the expense of others.

§ 2

If one continues the analogy to an engineering drawing, then a representation easily comes to mind in which this comparison is quite adequate. If one is considering a mountain, for example, then one imagines a horizontal plane passing through the base of the mountain. One further imagines lines passing through each point on the mountain and continuing to the plane; then the position of each point is where these lines meet the plane. In this manner one obtains a representation of the mountain, and it is only necessary to know the elevation of the points in order to obtain the true positions, since the plan view already gives the horizontal positions. For completeness sake profile drawings are included with the plan drawing in order to visualize the shape of the mountain heights. The more regular the mountain, the fewer profiles are required. If the mountain, for example, were a slice from a sphere, one profile alone would be sufficient.

§ 3

Now there is no reason why, say, Europe cannot be considered such as a mountain, with a base at the sea near Cape St. Vincent, at the mouth of the Nile, and at the northern tip of Norway. A plane surface can be thought to pass through these three points, and perpendiculars can be drawn through all places to give their positions. The profile can easily be drawn with a compass and the map of Europe is thus likened to the plan view of a mountain.

§ 4

The projection method just described is the so-called orthographic, and its properties have been known for a long time. It is thereby only possible to present at most one half of the earth's surface, and it has, as far as I am aware, never been used for maps of individual countries. However, astronomers have used it often for sketches of the stars. It has also been used for planetaria of the heavens, since the rising and setting of the sun and stars at various latitudes is easily visualized, and it can equally well be used for ready determination of the sine of the sun's altitude at every hour and for any latitude. It is only necessary to display the sphere orthographically on the surface of the Colurus Solstitiorum. The usual appearance of circles of the sphere on this projection is that of ellipses, though occasionally they are straight lines or circles.

§ 5

The notion that a terrestrial map should be viewed as a plan view of the earth's surface has not been binding, and perspective sketches have been employed instead of plan views. Here the earth's surface is drawn as it would be seen by the eye from a particular viewing point. Professor Karsten has recently advanced this portion of applied analysis by publishing a general analytical theory on the subject in Volume 5 of the Bavarian Proceedings. It is possible to choose innumerable points as the position of the eye from which to view the earth perspectively, but three advantageous points are emphasized. In one instance the eye is placed infinitely far from the globe, and this yields the aforementioned orthographic projection. In a second instance the point is taken somewhere on the surface of the earth, and this method of projection is called stereographic, presumably because of the lack of a more specific expression. Finally, the eye is taken at the midpoint of the earth, and this method of projection, since no other name is known to me, will be called the central projection.

§ 6

The stereographic method of projection contains many advantages since all circles of the sphere appear as either straight lines or as circles and since all angles on the surface of the sphere retain their size on the projection. It is therefore commonly used for constructions of the entire earth's surface and for regions of the world, as well as heavenly globes; in particular Mr. Hase has introduced what he calls the horizontal stereographic projection for particular regions, in which the eye is positioned at the nadir of the midpoint of the country to be projected. This is the eighth method described by Varenius. Mr. Kästner presented an analytical theory covering this at the Göttingen Society of Sciences, which has appeared in print as his dissertation in mathematics and physics. Mr. Hase used it especially because the separation of places from each other can still be measured fairly accurately on such maps, and the shape of countries is still fairly well preserved. In the delineation of hemispheres of the earth the eye is usually placed at the 90th and 270th degrees of the equator for the respective drawing of the old and new world. If one wants instead to make the position of the polar lands more available, then the eye is placed at the poles. This is the reason why the dual hemispheres of Graf von Redern, published by the Royal Academy of Sciences in Berlin, have the poles in the center; in this manner the positioning of the poorly known Southern regions becomes more apparent. Additionally, countries on the stereographic projection become larger the further they are from the center, while distances increase as the tangents of half the separation from the center.

§ 7

The central projection has the advantage that all spherical great circles are represented on it by straight lines. The small circles are in exceptional circumstances shown as circles but are otherwise always represented by conic sections. This method of composition therefore has the advantage that all places that lie on a straight line are on a great circle on the surface of the earth. I am not aware of any maps drawn in this manner, unless one includes those drawn on sundials where this construction occurs. Charts of the heavens, on the other hand, are advantageously drawn in this manner. Dopplemayer has thus made the whole heaven presentable on six plates, and with commendable accuracy. On the central projection less than a hemisphere can be represented because separations increase from the midpoint as the tangents. The sizes of the countries are exaggerated and shapes are noticeably disfigured.

§ 8

The three perspective compositions cited therefore have their advantages and disadvantages, and none satisfies all of the conditions cited in (§ 1). In particular, the condition that the sizes of the countries maintain their true relations is not found in any, and the conditions concerning measurement of the separation of places either requires restrictions or special constructions. This has been noted by Richmann in the treatise: "de perficiendis mappis geographicis, imprimis universalibus, per idoneas Scalas metiendis distantiis inservientes," published in 1751 in the thirteenth volume of the Petersburg Commentaries. The observations made by Richmann in this treatise are well taken, but require considerable extension and further improvement.

§ 9

Compositions for quite special purposes are required for the art of navigation. This has given a special form to nautical charts, which seem to have achieved perfection since the time of Mercator. Certainly these demonstrate that one cannot consider the perspective constructions as the only or the chief basis for the presentation of maps. Since not all conditions can be achieved simultaneously, it is adequate when a map satisfies those conditions necessary for its purpose. It would be advantageous if a composition could satisfy several requirements, and satisfy these exactly. For example, it is not apparent to what purposes the cylindrical map constructed by Bellin should serve, since it does not completely satisfy any particular condition.

§ 10

The elliptical figure of the earth is hardly sufficiently different from that of a sphere that one must take note of it in the composition of maps. Nevertheless Professor Lowitz has computed and presented, in the writings of the former Cosmological Society and in the German Staatsgeographicus, a representation of the spheroidal earth on which all angles preserve their size, as happens on the stereographic representation of the sphere and on Mercator's nautical charts. The location of countries and places is also far from exactly determined, so that differences of a few miles that might appear when large spheroidal distances are encountered are hardly to be considered. Such differences would also need to be assumed to be known exactly, and this is hardly the case. It is therefore not surprising that the spherical shape is retained. It becomes more elliptical than the earth anyway with the press of the copper, and differentially because paper shrinks, and by different amounts in length and width during the drying process. The earth can become oblong or oblate depending on the press of the copper. On these grounds alone the spheroidal figure is more an object for computation than for drawing. We shall not neglect it completely, however, and in the sequel will consider what consequences can thereby be observed.

§ 11

Those conditions mentioned at the onset that are imperfectly, or not at all represented by maps, are the proportionality of distances, and of the sizes of countries. The latter can be obtained, and even in many ways. Since Richmann has hardly started on it, there is adequate room to elaborate on this subject. The first condition cannot be achieved at all in any general and exact manner, especially if one chooses to use the standard, equally divided, rulers on the maps. Constructions with which distances can be recovered on various projections are proper, but usually too tedious and cumbersome. The simplest method is that used with Mercator's nautical chart. But this yields the length of the path of the vessel, and the actual distance only in those exceptional cases in which the travel is along a meridian. Richmann treats the matter as a problem in computation rather than one of constructions, and, since I am not familiar with anything else of value on this subject, I have treated it as completely unsolved, and will communicate what I have discovered of value.

I. MAPS TO DETERMINE THE DISTANCES OF PLACES

§ 12

Let there be three places A, P, B on the surface of the sphere. Their distances are

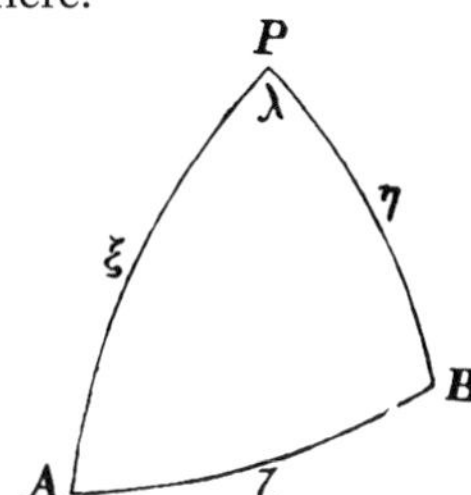

Fig. 1.

$$AP = \xi\,,$$
$$BP = \eta\,,$$
$$AB = \zeta$$

and the angle

$$APB = \lambda\,.$$

Here P can be the pole, and then one sees easily that λ is the difference in longitude of the two places A, B.

§ 13

Now let these places on the to-be-composed map be a, p, b. Let their distances be

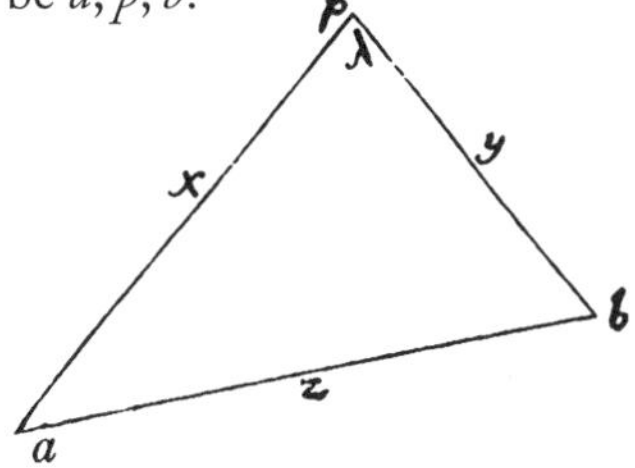

Fig. 2.

$$ap = x\,,$$
$$bp = y\,,$$
$$ab = z\,.$$

By this I mean the lengths of the straight lines ap, bp, ab.

§ 14

The three angles A, P, B on the spherical surface are together greater than 180 degrees, whereas the three plane angles a, p, b always equal 180 degrees. Therefore they cannot readily be compared with each other, nor can they be put into a proper relation with each other. I have therefore made the angle $apb = APB = \lambda$. But this fixes the position P in a manner which is not possible for the places A, B, and so I readily saw that all the places to be drawn on the map must have a relation to the point p. I was not deterred from investigating the consequences because I was free to make other stipulations if things did not proceed as desired.

§ 15

Now the trigonometry gave me the two equations

$$zz = xx + yy - 2\,xy\cos\lambda\,,\ \cos\zeta = \cos\xi\cos\eta + \sin\xi\sin\eta\cos\lambda\,.$$

In both of these the cosine of the angle λ occurs, and this was all as it should be, since it was adequate to note a variable relation between x, y, z and ξ, η, ζ. It was to be seen to what extent x, y, z were functions of ξ, η, ζ. And this question resolved itself into the extent to which the second equation could be transformed to be similar to the first equation.

§ 16

Now since ζ and z only occur once in the equation I began there, using the relation

$$\cos\zeta = 1 - 2\sin^2\tfrac{1}{2}\zeta$$

which gave the first transformation

$$2\sin^2\tfrac{1}{2}\zeta = 1 - \cos\xi\cos\eta - \sin\xi\sin\eta\cos\lambda\,.$$

Thus the first step toward the desired similarity of the two equations was achieved.

§ 17

For the other step I set

$$\cos\xi = 1 - 2\sin^2\tfrac{1}{2}\xi\,,\ \cos\eta = 1 - 2\sin^2\tfrac{1}{2}\eta$$

and obtained

$$2\sin^2\tfrac{1}{2}\zeta = 2\sin^2\tfrac{1}{2}\xi + 2\sin^2\tfrac{1}{2}\eta - 4\sin^2\tfrac{1}{2}\xi\sin^2\tfrac{1}{2}\eta - \sin\xi\sin\eta\cos\lambda\,.$$

or

$$\sin^2 \tfrac{1}{2}\zeta = \sin^2 \tfrac{1}{2}\xi + \sin^2 \tfrac{1}{2}\eta - 2\sin^2 \tfrac{1}{2}\xi \sin^2 \tfrac{1}{2}\eta - \tfrac{1}{2}\sin\xi \sin\eta \cos\lambda \,.$$

Here I saw that although it was possible to set

$$\sin \tfrac{1}{2}\zeta = z$$

it was not possible to set

$$\sin \tfrac{1}{2}\xi = x \,, \sin \tfrac{1}{2}\eta = y$$

in order to make the equations similar. The only recourse was to see if the term

$$2\sin^2 \tfrac{1}{2}\xi \sin^2 \tfrac{1}{2}\eta$$

could be removed from the equation

$$\sin^2 \tfrac{1}{2}\zeta = \sin^2 \tfrac{1}{2}\xi + \sin^2 \tfrac{1}{2}\eta - 2\sin^2 \tfrac{1}{2}\xi \sin^2 \tfrac{1}{2}\eta - \tfrac{1}{2}\sin\xi \sin\eta \cos\lambda \,.$$

This was achieved because

$$\sin^2 \tfrac{1}{2}\xi - \sin^2 \tfrac{1}{2}\xi \sin^2 \tfrac{1}{2}\eta = \sin^2 \tfrac{1}{2}\xi \cos^2 \tfrac{1}{2}\eta \,,$$
$$\sin^2 \tfrac{1}{2}\eta - \sin^2 \tfrac{1}{2}\eta \sin^2 \tfrac{1}{2}\xi = \sin^2 \tfrac{1}{2}\eta \cos^2 \tfrac{1}{2}\xi$$

I therefore obtained

$$\sin^2 \tfrac{1}{2}\zeta = \sin^2 \tfrac{1}{2}\xi \cos^2 \tfrac{1}{2}\eta + \sin^2 \tfrac{1}{2}\eta \cos^2 \tfrac{1}{2}\xi - \tfrac{1}{2}\sin\xi \sin\eta \cos\lambda \,.$$

§ 18

This equation appeared to be further removed from the desired similarity since the arcs ξ, η occur more frequently than originally, and are involved in all three components. Then to bring it closer to similarity I used

$$\sin\xi = 2\sin \tfrac{1}{2}\xi \cos \tfrac{1}{2}\xi \,, \sin\eta = 2\sin \tfrac{1}{2}\eta \cos \tfrac{1}{2}\eta \,,$$

in the last term. I thereby obtained

$$\sin^2 \tfrac{1}{2}\zeta = \sin^2 \tfrac{1}{2}\xi \cos^2 \tfrac{1}{2}\eta + \sin^2 \tfrac{1}{2}\eta \cos^2 \tfrac{1}{2}\xi$$
$$- 2\sin \tfrac{1}{2}\xi \cos \tfrac{1}{2}\xi \sin \tfrac{1}{2}\eta \cos \tfrac{1}{2}\eta \cos\lambda \,.$$

There was no recourse left, to prevent further complication, except to divide the entire equation by

$$\cos^2 \tfrac{1}{2}\xi \cos^2 \tfrac{1}{2}\eta$$

and luckily this worked, because the equation turned into

$$\frac{\sin^2 \frac{1}{2}\zeta}{\cos^2 \frac{1}{2}\xi \cos^2 \frac{1}{2}\eta} = \operatorname{tang}^2 \tfrac{1}{2}\xi + \operatorname{tang}^2 \tfrac{1}{2}\eta - 2 \operatorname{tang} \tfrac{1}{2}\xi \operatorname{tang} \tfrac{1}{2}\eta \cos\lambda$$

and therefore could be compared with

$$z^2 = x^2 + y^2 - 2\,xy \cos\lambda\,.$$

§ 19

I then obtained

$$x = \operatorname{tang} \tfrac{1}{2}\xi\,,$$
$$y = \operatorname{tang} \tfrac{1}{2}\eta\,,$$
$$z = \frac{\sin \frac{1}{2}\zeta}{\cos \frac{1}{2}\xi \cos \frac{1}{2}\eta} = \sin \tfrac{1}{2}\zeta \sec \tfrac{1}{2}\xi \sec \tfrac{1}{2}\eta\,.$$

Here x, y are similar functions of ξ, η. On the other hand z is not a function of only ζ but of ζ, ξ, η taken together. But it has the advantage of being independent of the angle λ, and as long as only this angle is changed, then the denominator $\cos \frac{1}{2}\xi \cos \frac{1}{2}\eta$ can be regarded as a constant coefficient. Finally, z depends on ξ exactly as it does on η.

§ 20

The value $2 \sin \frac{1}{2}\zeta$ is, however, the chord of the arc AB, or the chord of the distance of the two places A, B. When λ only varies, then z varies as this chord. This observation makes it possible to easily determine the relation between z and the chords.

§ 21

Then set the angle $\lambda = 0$, and then $\zeta = \xi - \eta$. Since ξ, η are given, the chord $\xi - \eta$ is also given and can easily be compared with

$$z = x - y = \text{tang}\ \tfrac{1}{2}\,\xi - \text{tang}\ \tfrac{1}{2}\,\eta\,.$$

§ 22

This same comparison can be made when $\lambda = 180°$. For one then has

$$\zeta = \xi + \eta\,,$$
$$z = x + y = \text{tang}\ \tfrac{1}{2}\,\xi + \text{tang}\ \tfrac{1}{2}\,\eta$$

and z is in relation to the chord of $(\xi + \eta)$.

§ 23

We therefore have for the comparison of the two figures

$$(x - y) : \text{chord.}\ (\xi - \eta) = z : \text{chord.}\ \zeta\,,$$

or also

$$(x + y) : \text{chord.}\ (\xi + \eta) = z : \text{chord.}\ \zeta\,.$$

§ 24

If therefore a chord distance is entered on circular dividers then the distance x can be found in either of the following two ways: 1) Enter the distance $x - y$ on the arc $\xi - \eta$ to give the circular divider its setting. Then enter the distance z to obtain the length of ζ. 2) Alternately, enter the distance $x + y$ on the arc $\xi + \eta$. This latter procedure is more reliable, because $x - y$ is often very small, and can even be zero.

§ 25

In this computation the angles at P, p have preserved their sizes and the distances are taken as

$$x = \text{tang}\ \tfrac{1}{2}\xi\ , y = \text{tang}\ \tfrac{1}{2}\eta$$

and the designation of the maps is therefore stereographic. This is all the more fortunate since there are many planispheres, of both the earth and the heavens, drawn in this manner and with the pole as the mid-point. Another of the many attractive properties of this projection is therefore that the distances can so easily be obtained from chords using circular dividers, whenever the proportion between lines and chords can be established.

§ 26

Such a planisphere is represented in the third figure; p is the pole and a, b are two places whose distance is to be obtained. The parallel of b passes through c, γ, and one sees that $ac = 30°$, and $a\gamma = 110°$. The chord entries for the circular dividers

Fig. 3.

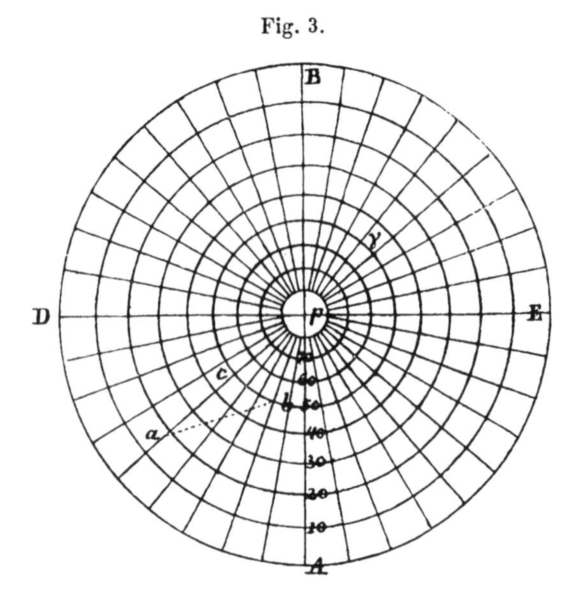

are therefore either 30 degrees or 110 degrees. One then enters *ab* and finds $39\frac{1}{2}$ degrees for the distance of the two places *a*, *b*.

§ 27

On the same method of projection let *Q* be one point, and let the points *A*, *B*, *C*, *D*, …, *M* all lie on a latitude circle then the lines *QA*, *QB*, *QC*, …, *QM* are proportional to the chords of the distances of the places *A*, *B*, *C*, …, *M* from *Q* (§ 20).

Fig. 4.

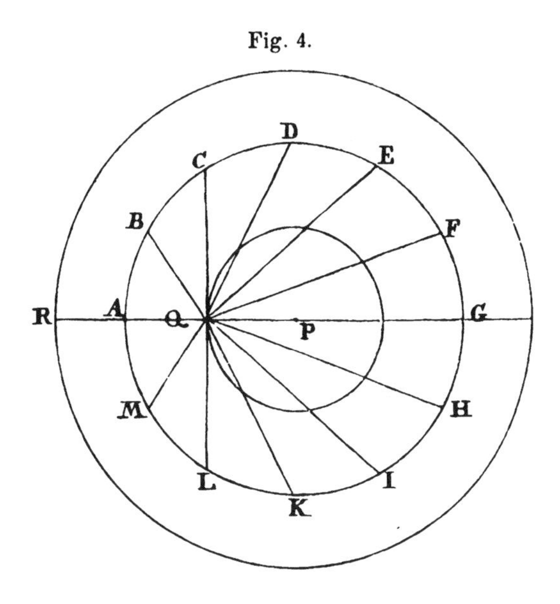

§ 28

Now if the places *A*, *B*, *C*, …, *M* are evenly distributed on the circle the famous rule of Cotes applies. The lines *QA*, *QB*, *QC*, …, *QM* are the coefficients of the binomial expression $a^m \pm b^m$.

Since these lines are in proportion to the chords of the distances of the places *A*, *B*, *C*, …, *M* from *Q*, one sees that Cotes' rule can be applied with minor modifications to the surface of a sphere, when it is stipulated that the point *Q* does not lie on the surface of the circle along which the points *A*, *B*, *C*, …, *M* are evenly distributed. But this is only noted in passing.

§ 29

If PQ is now taken to represent the separation of the pole from the zenith, and the circle $ADGK$ the latitude circle of the sun or a star, then the lines QA, QB, QC, ..., QM are proportional to the chords of the separation of the sun or the star from the zenith. For example, set the declination of the sun RA to 20°, set the polar height at $RQ = 50°$, then the separation of the latitude circle from the zenith $AQ = 30°$, and on the northern side $QG = 110°$. Then if on the chord lines of the proportional circle the distance AQ is set to the 30th, or QG set to the 110th degree, then the instrument has its proper opening, and if one enters the distance QB, QC, ..., QF for an arbitrary hour, then one will, in a very easy manner, find how much the sun or the star departs from the zenith at every hour. The height of the sun or the star above the horizon is then given by a simple subtraction of 90 degrees.

§ 30

If one does not have circular dividers at hand, then one must construct a circle, on which, for the first example, AQ is a chord of 30° or QG a chord of 110°, and then subdivide this circle into degrees.

§ 31

Or else one draws a scale on which AQ represents the sine of half of 30° or 15°, or QG the sine of half of 110° or 55°, then the lines QB, QC, QD, ..., QF on this scale will give the sines of half of the distances of the sun or the star from the zenith, and these can be referred to tables of sines and are easily doubled. This procedure is more troublesome than the use of circular dividers, but it is much more exact because the chord lines on circular dividers are seldom subdivided finer than whole degrees.

§ 32

After having obtained these results from the analytical formulae, their simplicity encouraged me to investigate whether these matters could not equally easily be obtained synthetically. To this end I composed the fifth figure, an orthographic representation of the sphere with the viewing point between the pole and the equator. The surface $DABa$ therefore represents the equator. P is the pole, lying on the front side of the sphere, p is the pole on the back side. $FPfp$ and $EPep$ are two meridians, M and N two points on them, and NR the arc through N parallel to the equator. Now the points M, N, R are supposed to be imagined as being projected stereographically onto the plane of the equator. For this purpose I drew CE, CF and these lines represent the projection of the longitude circles Pep and PFp. I also drew pM, pN, pR and these yielded the points m, n, r which therefore represent the stereographic construction of the points M, N, R. Now it is to be proven that

Fig. 5.

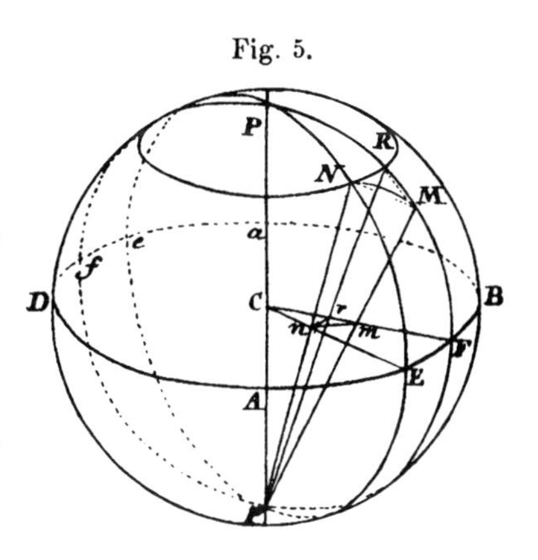

$$nm : rm = \text{chord } NM : \text{chord } RM\,.$$

§ 33

For this proof I will employ a theorem already given by Pappus. The two diameters Pp, CE of the circle Pep (see figure 6) are supposed perpendicular. Draw pE and an arbitrary chord pN from p, then

$$pP : pN = pn : pC\,.$$

This follows directly from the similarity of the two right-angled triangles pCn, pNp. Consequently

$$pN \cdot pn = pP \cdot pC = pE^2\,.$$

That is to say, the product of pN and pn is a constant, no matter how much the angle PpN is varied.

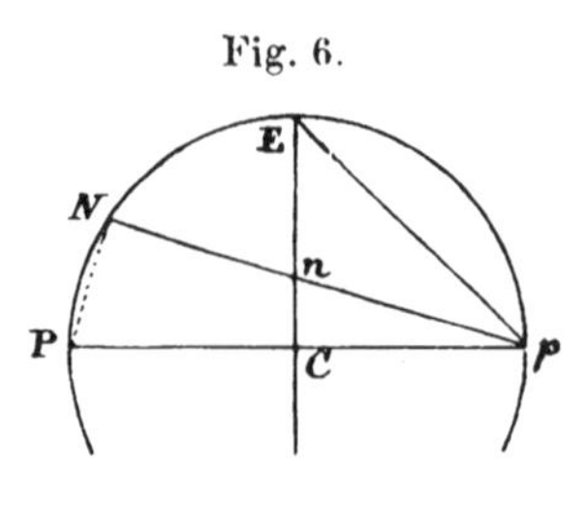

Fig. 6.

§ 34

As a consequence of this lemma the relations

$$pm \cdot pM = pr : pR = pn \cdot pN$$

hold in the fifth figure, and therefore

$$pm : pr = pR : pM\,.$$

This shows the triangles pmr, pRM to be similar, because they have the angle RpM in common, and then

$$pm : mr = pR : RM\,.$$

§ 35

Furthermore, since the points N, R lie on the same latitude circle,

$$pN = pR\,,$$
$$pn = pr\,.$$

The former analogy

$$pm : pr = pR : pM$$

becomes

$$pm : pn = pN : pM\,.$$

Thus the triangles pnm, pMN, because of the common angle at p, are similar to each other, and also

$$pm : mn = pN : NM\,,$$

or

$$pm : mn = pR : NM\,.$$

Since it was also found that

$$pm : mr = pR : RM\,,$$

these two analogies lead to

$$pm : pR = mn : NM = mr : RM$$

so that

$$mn : mr = NM : RM\,,$$

which was to be proven.

§ 36

For the representation of smaller portions of the earth's surface, e.g., for individual countries, Hase also used his oblique stereographic projection attempting thereby to obtain the advantage of being able to measure the separation of places with a simple ruler with more exactness than on any other representation. On the oblique projection the eye is at the nadir point of the center p of the map (figure 2). From (§ 19)

$$z = \frac{\sin\frac{1}{2}\zeta}{\cos\frac{1}{2}\xi\,\cos\frac{1}{2}\eta}\,,$$

so that

$$\sin\tfrac{1}{2}\zeta = z\cos\tfrac{1}{2}\xi\,\cos\tfrac{1}{2}\eta\,,$$

and

$$\frac{1}{2}\zeta = z\cos\tfrac{1}{2}\xi\cos\tfrac{1}{2}\eta + \tfrac{1}{6}z^3\cos^3\tfrac{1}{2}\xi\cos^3\tfrac{1}{2}\eta + \text{etc.}$$
$$= z\cos\tfrac{1}{2}\xi\cos\tfrac{1}{2}\eta\,(1 + \tfrac{1}{6}z^2\cos^2\tfrac{1}{2}\xi\cos^2\tfrac{1}{2}\eta + \text{etc.})\,.$$

Or since

$$\cos\tfrac{1}{2}\xi = 1 - 2\sin^2\tfrac{1}{4}\xi,$$
$$\cos\tfrac{1}{2}\eta = 1 - 2\sin^2\tfrac{1}{4}\eta,$$

so that

$$\tfrac{1}{2}\zeta = z\,(1 - 2\sin^2\tfrac{1}{4}\xi)\,(1 - 2\sin^2\tfrac{1}{4}\eta)$$
$$[1 + \tfrac{1}{6}z^2\,(1 - 2\sin^2\tfrac{1}{4}\xi)^2\,(1 - 2\sin^2\tfrac{1}{4}\eta)^2 + \text{etc.}]$$

or when fourth and higher dignitaries are neglected

$$\tfrac{1}{2}\zeta = z\,(1 - 2\sin^2\tfrac{1}{4}\xi - 2\sin^2\tfrac{1}{4}\eta + \tfrac{1}{6}z^2 + \text{etc.})\,.$$

§ 37

This formula shows that the error of a simple measurement begins to become noticeable when ξ, η are separated by several degrees. The error depends equally on ξ and η and becomes largest when one chooses points on the map farthest from the center. We can therefore set $\xi = \eta = 5, 10, 15, 20$ degrees in sequence. The error however also depends on z^2, and decreases when z increases. For a given value of ξ, η the distance z can range from 0 to tang $\frac{1}{2}\xi$ + tang $\frac{1}{2}\eta$ while the angle λ ranges from 0 to 180 degrees.

§ 38

In order to maintain these limits we want to set

$$z = n\,(\text{tang}\,\tfrac{1}{2}\xi + \text{tang}\,\tfrac{1}{2}\eta)\,.$$

For $\xi = \eta$ this yields

$$z = 2\,n\,\text{tang}\,\tfrac{1}{2}\xi\,.$$

Hence we obtain

$\xi = 5^\circ$	$\frac{1}{2}\zeta = 0{,}0873219\, n\,(1 - 0{,}0019036 + 0{,}0012708\, n^2)$
$= 10^\circ$	$= 0{,}1749773\, n\,(1 - 0{,}0076106 + 0{,}0051028\, n^2)$
$= 15^\circ$	$= 0{,}2633050\, n\,(1 - 0{,}0171103 + 0{,}0115549\, n^2)$
$= 20^\circ$	$= 0{,}3526540\, n\,(1 - 0{,}0303845 + 0{,}0207275\, n^2)$
etc.	etc.

§ 39

The coefficients enclosed by brackets determine the error relative to the total distance, because they indicate the proportion by which the distance must be diminished. They depend on n^2, and the proportionate reduction for $n = 1$ is approximately $\frac{2}{3}$. When $\xi = \eta = 20^\circ$ then the error is at most one mile in 30, but at least one mile in 100.

§ 40

If one instead wants to determine the absolute error it is necessary to take into account the entire expression. This attains a maximum when $\eta = \sqrt{\frac{1}{2}}$, when $\lambda = 90^\circ$. In this case one finds

$\xi = \eta = 5^\circ$	$\frac{1}{2}\zeta = 0{,}0\,617\,459 - 0{,}0\,000\,783$
$= 10^\circ$	$= 0{,}1\,237\,277 - 0{,}0\,000\,271$
$= 15^\circ$	$= 0{,}1\,861\,817 - 0{,}0\,021\,099$
$= 20^\circ$	$= 0{,}2\,493\,640 - 0{,}0\,049\,922$

or in degrees, minutes, and seconds

$\xi = \eta = 5^\circ$	$\frac{1}{2}\zeta = 3^\circ$	$32'$	$16''$	$-$	$0'$	$16''$
$= 10^\circ$	$= 7$	5	21	$-$	2	9
$= 15^\circ$	$= 10$	40	3	$-$	7	15
$= 20^\circ$	$= 14$	17	15	$-$	17	10

One thereby sees that if one does not wish to err by minutes then one cannot let ξ and η exceed 5°, and the map cannot extend over more than 10°.

§ 41

The errors given here are of course somewhat too small because of the neglected terms (§ 36). To determine them exactly, in the case for which $\lambda = 90^\circ$ and $\xi = \eta$ one uses

$$\cos \zeta = \cos^2 \xi$$

and this yields

$\xi = \eta$	$\frac{1}{2}\zeta$			$z - \frac{1}{2}\zeta$	
$= 5^\circ$	$= 3^\circ$	$32'$	$0''$	$= 0'$	$16''$
$= 10^\circ$	$= 7$	3	11	$= 2$	10
$= 15^\circ$	$= 10$	32	43	$= 7$	20
$= 20^\circ$	$= 13$	59	43	$= 17$	32

II. DISTANCES OF PLACES ON THE CENTRAL PROJECTION

§ 42

For the central projection the eye is placed in the middle of the globe, as was already mentioned above (§ 5). The eye therefore sees all points of the spherical surface in their true position. These positions however are altered by the projection. The largest circles of the sphere appear as straight lines, and the angles are replaced by their tangents.

§ 43

Henceforth let C be the point of tangency of the plane with the globe and let CR be the radius of the globe. If one completes the square $ARBDE$ so that $AR = RB = CR$ and with CR perpendicular to AB, then one sixth of the globe can be projected onto this face because it represents one side of a cube circumscribing the sphere. This is how Doppelmayer represented the heavens on six plates, as cited above (§ 7).

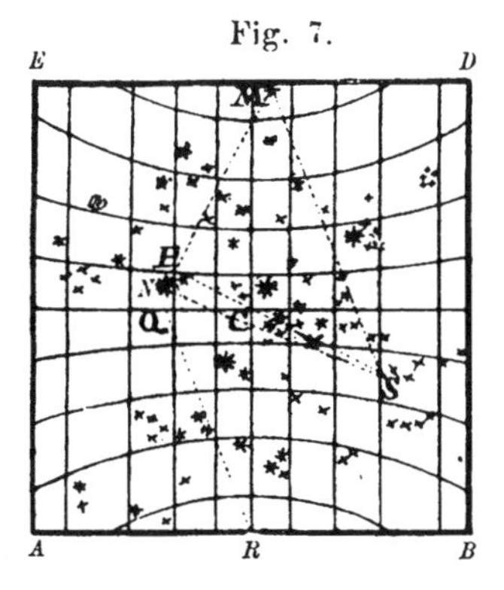

§ 44

Now let M, N be two stars (or two places if one is mapping the earth's surface). Draw a straight line MN through these, and it will represent a great circle of the sphere passing through the two stars or places. To determine the separation of these places it is only necessary to know the angle, which separates them at the center of the sphere or cube. For this purpose draw the perpendicular to MN through C to obtain PCS and find the distance of P from the eye. Extend CP from C to Q and RQ will be the requisite distance because CR is the distance of the eye from the viewing plane and the line from the eye to C intersects CP at right angles at C, here indicated by QCR. Now extend QR from P to S; then S is the point to which one must draw the lines NS, MS to obtain the angle

NSM which represents the separation of *M*, *N* at the eye or in the middle of the sphere or cube. Measurement of this angle gives the separation of the stars *M*, *N* in degrees, minutes, etc. This method permits very simple measurement of the separation of the stars on the six Doppelmayer plates.

III. CONSTRUCTION TO DETERMINE THE DISTANCES

§ 45

On the orthographic projection the separation of places can be determined through a simple construction, as is apparent from what has already been said (§ 3). This can also be found in Varenius who refers to the original discoverer Maurolycus. Since this presents no difficulties I will not tarry here but will instead return to the formula (§ 18)

$$\sin^2 \tfrac{1}{2}\zeta = \sin^2 \tfrac{1}{2}\xi \cos^2 \tfrac{1}{2}\eta + \cos^2 \tfrac{1}{2}\xi \sin^2 \tfrac{1}{2}\eta$$
$$- 2 \sin \tfrac{1}{2}\xi \cos \tfrac{1}{2}\eta \cos \tfrac{1}{2}\xi \sin \tfrac{1}{2}\eta \cos\lambda$$

and will use this to derive an alternate construction.

§ 46

To this end take this formula in quadruplicate:

$$(2 \sin \tfrac{1}{2}\zeta)^2 = (2 \sin \tfrac{1}{2}\xi \cos \eta)^2 + (2 \cos \tfrac{1}{2}\xi \sin \tfrac{1}{2}\eta)^2$$
$$- 8 \sin \tfrac{1}{2}\xi \cos \tfrac{1}{2}\eta \cos \tfrac{1}{2}\xi \sin \tfrac{1}{2}\eta \cos\lambda\,.$$

Since

$$2 \sin \tfrac{1}{2}\xi \cos \tfrac{1}{2}\eta = \sin \frac{\xi+\eta}{2} + \sin \frac{\xi-\eta}{2},$$
$$2 \cos \tfrac{1}{2}\xi \sin \tfrac{1}{2}\eta = \sin \frac{\xi+\eta}{2} - \sin \frac{\xi-\eta}{2},$$
$$2 \sin \tfrac{1}{2}\zeta = \text{chord.}\,\zeta$$

we obtain

$$(\text{chord}\,\zeta)^2 = \left[\sin \frac{\xi+\eta}{2} + \sin \frac{\xi-\eta}{2}\right]^2$$
$$+ \left[\sin \frac{\xi+\eta}{2} - \sin \frac{\xi-\eta}{2}\right]^2$$
$$-2\left[\sin \frac{\xi+\eta}{2} + \sin \frac{\xi-\eta}{2}\right] \cdot \left[\sin \frac{\xi+\eta}{2} - \sin \frac{\xi-\eta}{2}\right] \cdot \cos\lambda\,.$$

This formula can, as in (§ 18) be compared with

$$z^2 = x^2 + y^2 - 2xy\cos\lambda\,,$$

so that

$$x = \sin\frac{\xi+\eta}{2} + \sin\frac{\xi-\eta}{2}\,,$$
$$y = \sin\frac{\xi+\eta}{2} - \sin\frac{\xi-\eta}{2}\,,$$
$$z = \text{chord.}\,\zeta.$$

Everything depends on the sines of the half sum and half difference of the colatitudes. These can be taken from tables and used to construct figure 2, which gives the side z to the same scale, and therefore the chord separation of the two places.

IV. MORE GENERAL METHOD TO REPRESENT THE SPHERICAL SURFACE SO THAT ALL ANGLES PRESERVE THEIR SIZE

§ 47

Stereographic representations of the spherical surface, as well as Mercator's nautical charts, have the peculiarity that all angles maintain the sizes that they have on the surface of the globe. This yields the greatest similarity that any plane figure can have with one drawn on the surface of a sphere. The question has not been asked whether this property occurs only in the two methods of representation mentioned or whether these two representations, so different in appearances, can be made to approach each other through intermediate stages. Mercator represents the meridians as parallel lines, perpendicular to the equator, subdivided according to the logarithms of the cotangents of half the colatitudes. The equator itself is divided into 360 equal parts, as many as there are degrees. The angle of intersection of the meridians is then $= 0$, because they are parallel. Contrastingly, in the polar case of the stereographic projection, the same straight meridians intersect at the proper angle (figure 3). Consequently, if there are stages intermediate to these two representations, they must be sought by allowing the angle of intersection of the meridians to be arbitrarily larger or smaller than its value on the surface of the sphere. This is the way in which I shall now proceed.

§ 48

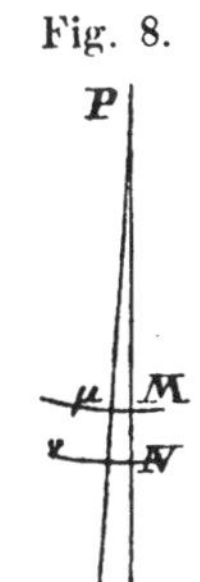

Let P be the pole; PM, $P\mu$ two meridians, and the angle $MP\mu$, infinitely small. The point M has ε as its colatitude, and N has $\varepsilon + d\varepsilon$. For the angle $MP\mu$ use $md\lambda$ where $d\lambda$ is the difference in longitude and m is the ratio of $MP\mu$ the true angle. The requirement is now that

$$\mu M : MN = d\lambda \sin \varepsilon : d\varepsilon$$

is supposed to hold, and the trapezoid $\mu MN\nu$ is to be similar to the one which it represents on the sphere. Let

$$PM = x$$
$$MN = dx\,,$$

then

$$M\mu = xm\,d\lambda\,,$$

from which

$$mx\,d\lambda : dx = d\lambda \sin\varepsilon : d\varepsilon\,.$$

It now follows that

$$\frac{dx}{x} = \frac{m\,d\varepsilon}{\sin\varepsilon}$$

and

$$\log x = m \log \operatorname{tang} \tfrac{1}{2}\varepsilon\,.$$

The constant can be omitted here and then $x = 1$ when $\varepsilon = 90^\circ$ no matter what the value of m. Consequently $x = (\operatorname{tang} \frac{1}{2}\varepsilon)^m$.

§ 49

If m is now set $= 1$, then $x = \operatorname{tang} \frac{1}{2}\varepsilon$, which is the case represented by the stereographic projection.

§ 50

For Mercator's projection $m = 0$. This yields only $x = 1$. But we can set $\varepsilon = 90^\circ - p$ and then

$$x = \operatorname{tang}^m (45 - \tfrac{1}{2}p)$$
$$= \left(\frac{1 - \operatorname{tang}\frac{1}{2}p}{1 + \operatorname{tang}\frac{1}{2}p}\right)^m,$$

which gives

$$x = (1 - m \operatorname{tang} \tfrac{1}{2} p + m \frac{m-1}{2} \operatorname{tang}^2 \tfrac{1}{2} p - \text{etc.})$$
$$\times (1 - m \operatorname{tang} \tfrac{1}{2} p + m \frac{m+1}{2} \operatorname{tang}^2 \tfrac{1}{2} p - \text{etc.}),$$

so that for $m = 0$

$$\frac{1-x}{2m} = \operatorname{tang} \tfrac{1}{2} p + \tfrac{1}{3} \operatorname{tang}^3 \tfrac{1}{2} p + \tfrac{1}{5} \operatorname{tang}^5 \tfrac{1}{2} p + \tfrac{1}{7} \operatorname{tang}^7 \tfrac{1}{2} p + \text{etc.}$$
$$= \tfrac{1}{2} \log \operatorname{cotg} \tfrac{1}{2} \varepsilon$$

is obtained. Here $\frac{1-x}{m}$ represents the degrees computed from the equator and these increase in proportion to $\log \operatorname{cotg} \frac{1}{2} \varepsilon$.

§ 51

The choice of m in the general formula

$$x = (\operatorname{tang} \tfrac{1}{2} \varepsilon)^m$$

is arbitrary, so we can impose a further condition. In the trapezoid $\mu MN\nu$ of figure 8 the proportion νN to MN as well as μM to MN is to be as on the surface of the sphere. This cannot be obtained for all colatitudes, but is to be determined for a specific equatorial height E. Now

$$N\nu = m\, d\lambda\, (x + dx)$$

and this arc on the spherical surface is

$$d\lambda \sin (\varepsilon + d\varepsilon) = d\lambda\, (\sin \varepsilon + \cos \varepsilon\, d\varepsilon)\,.$$

This gives

$$d\varepsilon : dx = d\lambda\, (\sin \varepsilon + \cos \varepsilon\, d\varepsilon) : m\, d\lambda\, (x + dx)\,.$$

From this it follows that

$$d\varepsilon \cdot m\, (x + dx) = dx \sin \varepsilon + dx \cos \varepsilon\, d\varepsilon\,,$$
$$\frac{mx}{dx} + m = \frac{\sin \varepsilon}{d\varepsilon} + \cos \varepsilon$$

But from (§ 48)

$$\frac{m\,x}{d\,x} = \frac{\sin\varepsilon}{d\varepsilon}$$

or

$$\frac{\sin\varepsilon}{d\varepsilon} + m = \frac{\sin\varepsilon}{d\varepsilon} + \cos\varepsilon,$$

which gives

$$m = \cos\varepsilon\,,$$

and so for the specific equatorial height E

$$m = \cos E\,.$$

And hence the formula

$$x = (\text{tang}\,\tfrac{1}{2}\,\varepsilon)^{\cos E}$$

is thereby in the form which makes the trapezoid $\mu M \mathrm{N} \nu$ completely similar to its spherical image at the equatorial height E. The proportionality to the lines on the spherical surface holds not only for μM relative to MN but also for $N\nu$.

§ 52

This completely true proportionality holds exactly only for equatorial height = E, but it departs from the true value as little as possible on either side; it is in other words a true minimum. As a consequence, if one wishes to construct a map of Europe, for example, the equatorial height E should be chosen to be a midlatitude of Europe. E will therefore be about 40°. It is convenient to take

$$E = 41^\circ\, 24'\, 35''$$

and then

$$\cos E = \tfrac{3}{4}.$$

For a map of Europe the formula becomes

$$\text{Log}\, x = \tfrac{3}{4}\,\text{Log tang}\,\tfrac{1}{2}\,\varepsilon\,.$$

And this yields

ε	x	Diff.
10	0,16 087	0,16 087
20	0,27 211	0,11 124
30	0,37 243	0,10 032
40	0,46 860	0,09 617
50	0,56 429	0,09 569
60	0,66 234	0,09 805
70	0,76 546	0,10 312
80	0,87 672	0,11 126
90	1,00 000	0,12 328

It is clear that a map of Europe only requires from the 20th to the 60th degrees. And since $m = \cos E = \frac{3}{4}$, the degrees of longitude are smaller by $\frac{1}{4}$ so that 30 compass degrees represent 40 degrees of longitude. The drawing is represented in figure 9.

Fig. 9.

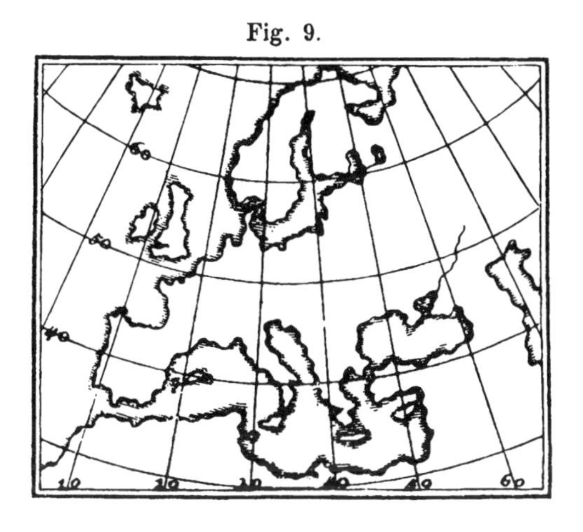

§ 53

The value for m can also be determined so that the degrees of two given parallels maintain their correct proportions. Let the two parallel circles have colatitudes a, b; then by the condition it is necessary that

$$(\text{tang } \tfrac{1}{2} a)^m : (\text{tang } \tfrac{1}{2} b)^m = \sin a : \sin b$$

From this it follows that

$$m = \frac{\text{Log sin } a - \text{Log sin } b}{\text{Log tang } \frac{1}{2} a - \text{Log tang } \frac{1}{2} b}.$$

§ 54

For the European case, for example, one takes the equatorial heights

$$a = 60^\circ, b = 20^\circ,$$

then

$$\begin{array}{rl} \text{Log tang } \frac{1}{2} a = & 9{,}7\ 614\ 394 \\ \text{Log tang } \frac{1}{2} b = & \underline{9{,}2\ 463\ 188} \\ & 0{,}5\ 151\ 206 \end{array} \qquad \begin{array}{rl} \text{Log sin } a = & 9{,}9\ 375\ 306 \\ \text{Log sin } b = & \underline{9{,}5\ 340\ 517} \\ & 0{,}4\ 034\ 789 \end{array}$$

and therefore

$$m = \frac{0{,}4\ 034\ 789}{0{,}5\ 151\ 206} = 0{,}78\ 327.$$

This is the value of the cosine of 38° 26′, so that $E = 38^\circ\ 26'$.

One can choose this value for E and it is clear that the condition specified here and the earlier (§ 51) condition can be satisfied simultaneously. One can thus choose E and can then find, for every colatitude a, another b such that the degrees of the parallel circles maintain their true proportions for both. For example, if one takes, as before (§ 52),

$$\text{Cos } E = \tfrac{3}{4}$$

and assumes $a = 60^\circ$, then one finds $b = 24^\circ\ 56'$. For

$$\begin{array}{rl} \text{Log tang } \frac{1}{2} a = & 9{,}7\ 614\ 394 \\ \text{Log tang } \frac{1}{2} b = & \underline{9{,}3\ 445\ 580} \\ & 0{,}4\ 168\ 814 \end{array} \qquad \begin{array}{rl} \text{Log sin } a = & 9{,}9\ 375\ 306 \\ \text{Log sin } b = & \underline{9{,}6\ 248\ 629} \\ & 0{,}3\ 126\ 677 \end{array}$$

from which

$$m = \frac{0{,}3\ 126\ 677}{0{,}4\ 168\ 814} = \tfrac{3}{4}.$$

§ 55

If one wanted to set $m = 1$, as is the case for the stereographic projection, or $m = 0$ as on Mercator's nautical charts, then in both cases one would obtain $a = b$. Consequently there are not two distinct parallels which have the same ratios as on the spherical surface for these methods of representation, in spite of the fact that there are innumerable such parallels as soon as m lies between 0 and 1.

§ 56

When one assumes $m < 1$ then the degrees of longitude are reduced in the ratio of 1 to m, and only 360 m degrees of a circle are required to represent the 360 degrees of longitude. If one takes, e.g., as before, $m = \frac{3}{4}$ then one only needs a sector of 270 degrees to represent the 360 degrees of longitude. If the entire northern or southern hemisphere is to be represented, and the beginning is to fit to the end, then the 270 degree sector can be rolled into a cone. This yields conical globes, on which it has long been the custom to represent hemispheres of the heavens. Here we have achieved such a representation in which all angles preserve their true magnitude, and consequently the star images have the largest possible similarity to the heavens. Zimmermann's well-known star cones do not have this advantage. The sector there is 300 degrees and the meridians are divided into 90 equal parts which makes 10 degrees of the equator equal to 13 degrees of the meridian so that the lengths of the latitudes are not proportional to the longitudes.

§ 57

For such star cones the value of m in the formula

$$x = (\text{tang}\ \tfrac{1}{2}\ \varepsilon)^m$$

is equally available, and can be chosen to satisfy particular conditions; so let us require that the first 45° be equal to the ones which follow, in order that the 45th parallel on the star cone divides the meridian into two equal parts. By this condition $x = \frac{1}{2}$ when $\varepsilon = 45°$ so that

$$\tfrac{1}{2} = (\text{tang}\ 22\ \tfrac{1}{2})^m\ ,$$

$$m = \frac{\text{Log}\ 2}{\text{Log tang}\ 67\frac{1}{2}°} = \frac{0{,}3\ 010\ 300}{0{,}3\ 827\ 757} = 0{,}78\ 643,$$

instead of which, since small differences are not worth consideration, and for more convenient computation one can take $m = \frac{4}{5}$ or, as in the previous case $m = \frac{3}{4}$, depending on whether one wants the cone to be more or less flat. If one wants the same flatness as those of Zimmermann then it is necessary to take $m = \frac{5}{6}$.

V. FURTHER EXTENSIONS OF THE SAME METHOD

§ 58

We now postulate that the meridians are to be circles which cross at both of the poles. If this takes place so that the angles of intersection maintain their magnitudes, then this is the stereographic method of representation, used for most planispheres of the earth and therefore long known. The parallel circles also appear circular here, and cross the meridians at right angles, so that the proportion of length to breadth is maintained throughout.

§ 59

This situation will be used to advantage by requiring that the meridians intersect at the poles in angles which are related to the true angles by the ratio of 1 to m. The question is now to sketch the parallels so that the proportions between the degrees of length and breadth, and thereby all angles, maintain their proper magnitude.

§ 60

Here the parallels as before remain circular and they intersect the circles representing meridians as before at right angles. And they are also drawn as on the stereographic projection. It remains only to determine the point through which each parallel is to be drawn.

§ 61

Let P, p be the two poles, Pp a meridian, here taken to be a straight line. Divide Pp into two equal pieces at A, and draw the perpendicular DAB through A; this is to represent the equator. Let B be the center of a circular arc PMp, which is

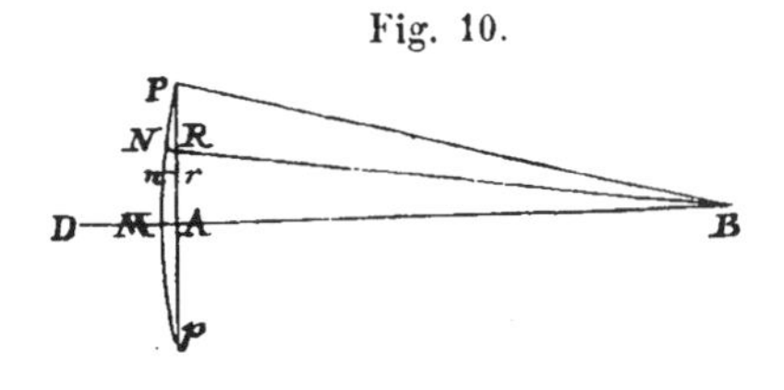

Fig. 10.

to represent a meridian separated from PAp by an infinitely small distance. The difference in longitude is to be $= \lambda$, so that the angle must be given by $MPA = m\lambda$. Set $AP = 1$, then $MA = \text{tang } \frac{1}{2}\, m\lambda$, or, since λ is infinitely small

$$MA = \tfrac{1}{2}\, m\lambda \,.$$

Furthermore $AB = \text{cotang } m\lambda$, or also because λ is infinitely small

$$AB = \frac{1}{m\lambda} \,.$$

Let the equatorial height of the point R be ε, and let

$$AR = x, - Rr = dx \,,$$

then

$$MB = NB = \frac{1}{m\lambda} + \tfrac{1}{2}\, m\lambda \,,$$

$$RB = \sqrt{\frac{1}{m^2\lambda^2} + x^2} = \frac{1}{m\lambda} + \tfrac{1}{2}\, x^2\, m\lambda$$

hence

$$NR = \tfrac{1}{2}\, m\lambda\, (1 - xx) \,.$$

Now, in order that the degrees of longitude have the correct relation to the degrees of latitude, one must have

$$NR : Rr = \lambda \sin \varepsilon : d\varepsilon$$

Then

$$- \tfrac{1}{2}\, m\lambda\, (1 - xx) : dx = \lambda \sin \varepsilon : d\varepsilon \,,$$

Whence

$$- \frac{md\varepsilon}{\sin \varepsilon} = \frac{2dx}{1 - xx}$$

follows. Set

$$x = \cos \varphi \,,$$

then

$$- \frac{md\varepsilon}{\sin \varepsilon} = \frac{2\delta\varphi}{\sin \varphi} \,,$$

which gives

$$(\text{tang } \tfrac{1}{2}\,\varepsilon)^m = \text{tang}^2\, \tfrac{1}{2}\,\varphi$$

from which, since

$$x = \frac{1 - \text{tang}^2\, \frac{1}{2}\,\varphi}{1 + \text{tang}^2\, \frac{1}{2}\,\varphi}$$

one obtains

$$x = \frac{1 - (\text{tang}^2\, \frac{1}{2}\,\varepsilon)^m}{1 + (\text{tang}^2\, \frac{1}{2}\,\varepsilon)^m} = 1 - \frac{2}{\text{cotg}^m\, \frac{1}{2}\,\varepsilon + 1}\,.$$

§ 62

Here one again has the option of choosing m as desired. What suggests itself most naturally is to take

$$m = \tfrac{1}{2}$$

Otherwise, with $m = 0$ this reduces to Mercator's drawing of nautical charts. And with $m = 1$ the well known stereographic projection reappears, with only half of the spherical surface represented inside the circular area of radius AP. With $m = \frac{1}{2}$, on the other hand, the entire spherical surface is represented there because the meridians intersect at the poles at the half angles, and the 360 degrees of length are reduced to 360 half or 180 whole degrees.

§ 63

To this objective I have taken $m = \frac{1}{2}$ and found that for

$\varepsilon =$	90°	$x =$	0,00 000	$\varepsilon =$	40°	$x =$	0,24 746
	80 °		0,04 383		30°		0,31 783
	70°		0,08 888		20°		0,40 856
	60°		0,13 648		10°		0,54 346
	50°		0,18 844				

This permits drawing of the parallels every 10 degrees, as presented in the 11th figure. They cross the meridians at right angles, and the meridians themselves intersect at both poles at angles which are half as large as the true ones. Finally at all places the proportion of degrees of length to degrees of breadth is correct, and all angles maintain their magnitude, the two poles being excepted since the angles here have half their size.

Fig. 11.

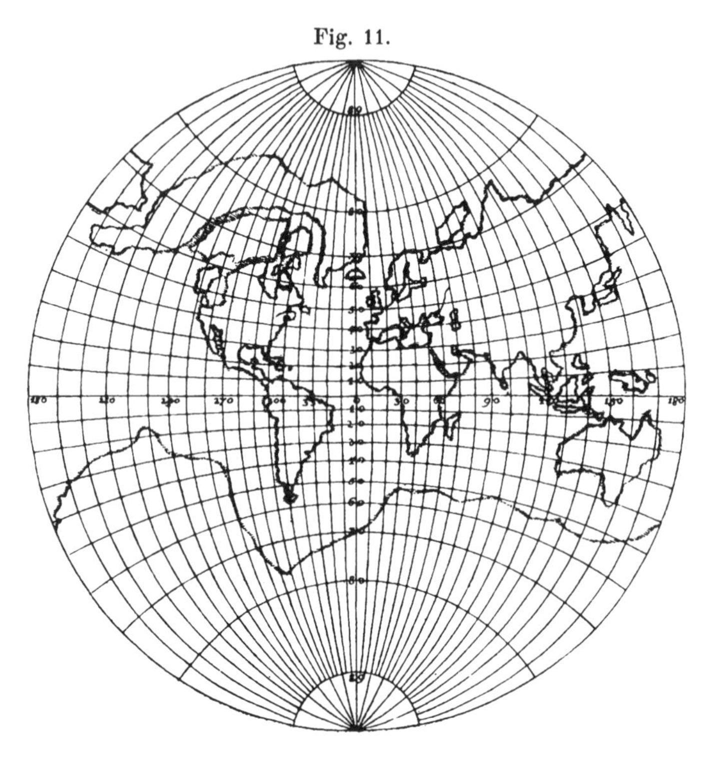

§ 64

It should also be noted that, in integrating the formula (§ 61)

$$\frac{m\,d\varepsilon}{\sin\varepsilon} = \frac{2\delta\varphi}{\sin\varphi}$$

whose integral is

$$\tfrac{1}{2}\,m\log\text{tang}\,\tfrac{1}{2}\,\varepsilon = \log\text{tang}\,\tfrac{1}{2}\,\varphi + \text{Const.}$$

the constant is still available. It can be established, for example by requiring that $x = 0$, and thereby $\varphi = 90^{\circ}$, when $\varepsilon = E$. This will give the more general formula

$$x = 1 - \frac{2}{(\text{tang } \frac{1}{2} E)^{m} \cdot (\text{cotg } \frac{1}{2} \varepsilon)^{m} + 1}$$

One thus obtains that the line which represents the equator in the 10th figure can now represent any arbitrary parallel. No matter what this value one has for the pole, at $\varepsilon = 0$,

$$x = +1$$

and for the pole at $\varepsilon = 180^{\circ}$,

$$x = -1$$

and both poles are therefore equidistant from the midpoint. The consequence is that the degrees of latitude are dissimilar on the two sides of the equator, and the drawing is more irregular than if one takes $E = 90^{\circ}$ and therefore $\text{tang } \frac{1}{2} E = 1$, as has been done for the figure and which is incomparably more satisfying.

VI. MOST GENERAL LECTURE ON THE SAME METHOD

§ 65

If the question is one of universally representing the spherical surface so that the meridians and parallels always intersect at right angles, and that the ratios of the degrees of latitude and of longitude are maintained for all locations, with the consequence that all angles preserve their magnitude, then the cases considered thus far constitute only a small proportion of the total possible cases with these attributes, for the curves may be represented not only by straight lines and arcs of circles, but by innumerable types of curved lines and still satisfy the stated requirements. This task, outlined here in full generality, if not more difficult than, is at least not easier than that of determining reciprocal trajectories, and is very closely related to that problem since several of the types of trajectories also appear as solutions to the present problem.

§ 66

It does not appear that one should expect a general solution to this problem, unless one is satisfied with differential and integral formulae or general specification of functions, since it is easy to predict that the solution will be expressed in the form of innumerable and unrelated types of curved lines. I have therefore, immediately as I found the underlying differential equations, turned to infinite series, and would have done this even if the complete and general integral of these differential equations had been more apparent. Often one must, lacking a method leading directly to the goal, find integrals by probing in the dark, and I include infinite series among such probing. It is clear that these can be summed whenever the differential is in fact integrable, and I have often encountered cases in which the series solution could be found more easily than the integral, and which then suggested the path by which this could have been found if one had known it in advance.

§ 67

In figure 12 let CA represent the abscissa; CE a perpendicular, parallel to the ordinate; EM the arc of a parallel of colatitude p, and LM the arc of a meridian of longitude λ.

Fig. 12.

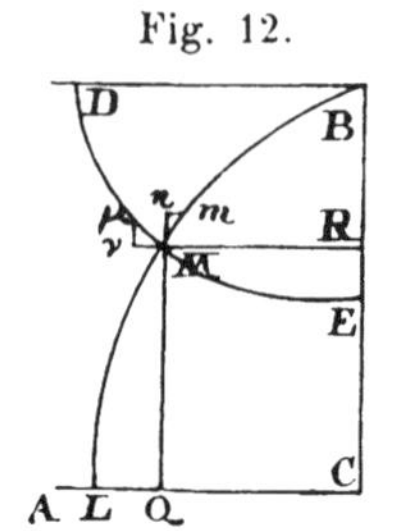

Draw the ordinate nMQ, and set

$$CQ = x\,,\; QM = y\,.$$

In addition let m be a point of colatitude $p + dp$ on the meridian, and let μ be a point of longitude $\lambda + d\lambda$ on the parallel. Construct nm, νM parallel to AC, and $\mu\nu$ parallel to CE; then the objective of the assignment requires that the angle $\mu Mm = 90° = \nu Mn$ consequently the right triangles $\nu M\mu$, nMm at ν, and n must be similar; furthermore μM to Mm must remain as the degrees of longitude to those of latitude; consequently

$$\mu M : Mm = \cos p\, d\lambda : dp$$

which can be replaced, because of the similarity of the triangles, by the two analogous relations

$$\nu M : Mn = \cos p\, d\lambda : dp\,,$$
$$\nu\mu : mn = \cos p\, d\lambda : dp\,,$$

which includes both conditions.

§ 68

Now x as well as y is a function of p and λ. Consequently if one differentiates x and y, then each differential has a portion multiplied by dp and a portion multiplied by $d\lambda$ and, more generally we can set

$$dy = M dp + m\, d\lambda$$
$$dx = N dp + n\, d\lambda$$

and regard M, m, N, n, as functions of p and λ. Now for any single meridian LM, the longitude λ, is constant, therefore $d\lambda = 0$. From the figure this yields

$$+ dy' = Mn = Mdp$$
$$- dx' = - mn = Ndp$$

Similarly, for any single parallel EM the latitude p is invariant, and thereby $dp = 0$. This yields

$$+ dy'' = \mu v = m d\lambda$$
$$+ dx'' = vM = n d\lambda$$

so that the total differentials are

$$dy = dy' + dy'' = Mn + \mu v\,,$$
$$dx = - dx' + dx'' = - mn + Mv$$

§ 69

Substituting these values into the two previous analogies, one obtains

$$+ n d\lambda : + Mdp = \cos p\, d\lambda : dp\,,$$
$$+ m d\lambda : - Ndp = \cos p\, d\lambda : dp\,.$$

And from this

$$+ M \cos p = n\,,$$
$$- N \cos p = m\,.$$

§ 70

Using these two equalities, two of the four functions M, m, N, n, in the equations of (§ 68)

$$dy = Mdp + m d\lambda\,,$$
$$dx = Ndp + n d\lambda$$

can be eliminated, and one obtains if, e.g., one eliminates M and n

$$dy = \frac{ndp}{\cos p} - N \cos p\, d\lambda,$$

where y is a function of the variables n, N obtained from x, and is consequently itself a function of x.

§ 71

As an example, on Mercator's design

$$x = \lambda \,,$$

consequently

$$dx = d\lambda \,.$$

This makes

$$N = 0 \,, n = 1 \,,$$

consequently

$$dy = \frac{dp}{\cos p} \,,$$

$$y = \log \operatorname{tang} (45^\circ + \tfrac{1}{2} p) \,.$$

§ 72

The same holds for the stereographic projection, where C is the midpoint and CA is the equator

$$x = \frac{\sin \lambda \cos p}{1 + \cos \lambda \cos p} \,,$$

consequently

$$dx = \frac{(\cos \lambda + \cos p) \cos p \, d\lambda - \sin p \sin \lambda \, dp}{(1 + \cos \lambda \cos p)^2} \,.$$

This makes

$$N = - \frac{\sin p \sin \lambda}{(1 + \cos \lambda \cos p)^2} \,,$$

$$n = + \frac{(\cos \lambda + \cos p) \cos p}{(1 + \cos \lambda \cos p)^2} \,,$$

consequently

$$dy = \frac{(\cos\lambda \cos p)\, dp + \sin p \cos \sin \lambda\, d\lambda}{(1 + \cos\lambda \cos p)^2},$$

$$y = \frac{\sin p}{1 + \cos\lambda \cos p}.$$

§ 73

In this manner x is determined by λ, p can be taken arbitrarily, and the corresponding differential equation for y can be found. And since one finds, in the same fashion ,

$$dx = -\frac{m}{\cos p}\, dp + M \cos p\, d\lambda$$

so one also has the choice of assuming an equation for y, and thereby finding the differential equation for x. It is also easily possible, however, that such an equation represents nothing, since it must be constructed so that, if one regards as variable only λ in $-\frac{m}{\cos p}$ and only p in $M \cos p$, these two expressions result in the same value upon differentiation, if the first is divided by $d\lambda$ and the second by dp. This is also valid for the first equation (§ 70)

$$dy = \frac{n\, dp}{\cos p} - N \cos p\, d\lambda.$$

One can also, starting from the differential relations

$$dy = M\, dp + m\, d\lambda\,,$$
$$dx = N\, dp + n\, d\lambda$$

using the equations

$$+ M \cos p = n$$
$$- N \cos p = m$$

eliminate M and N, and then one obtains

$$dy = \frac{n\, dp}{\cos p} + m\, d\lambda\,,$$

$$dx = \frac{m\, dp}{\cos p} + n\, d\lambda\,.$$

These equations now have the advantage that only one term is encumbered with dp and $\cos p$. This moved Mr. de la Grange, to whom I communicated the objective, to pursue additional investigations. For this purpose he set

$$\frac{dp}{\cos p} = d\mu$$

yielding

$$\mu = \log \text{tang}\,(45^{\circ} + \tfrac{1}{2}\,p)$$

and from which follows

$$dy = +\,n\,d\mu + m\,d\lambda\,,$$
$$dx = -\,m\,d\mu + n\,d\lambda\,,$$

consequently

$$dy\sqrt{-1} = +\,n\,d\mu\sqrt{-1} + m\sqrt{-1}\,d\lambda\,,$$
$$dx = +\,m\sqrt{-1}\,d\mu\sqrt{-1} + n\,d\lambda\,;$$

continuing from this one has

$$dx + dy\,\sqrt{-1} = (m\,\sqrt{-1} + n)\cdot(d\mu\,\sqrt{-1} + d\lambda)\,,$$
$$dx - dy\,\sqrt{-1} = (m\,\sqrt{-1} - n)\cdot(d\mu\,\sqrt{-1} - d\lambda)\,.$$

This shows generally that $m\sqrt{-1} + n$ must be a function of $\mu\sqrt{-1} + \lambda$ and similarly $m\sqrt{-1} - n$ must be a function of $\mu\sqrt{-1} - \lambda$. (See e.g., Bougainville, *Calcul Integral*, P. II, Chap. 16, Probl. 2).

Thereby

$$x + y\,\sqrt{-1} \text{ is a function of } \mu\,\sqrt{-1} + \lambda$$

and

$$x - y\,\sqrt{-1} \text{ is a function of } \mu\,\sqrt{-1} - \lambda\,.$$

The application of this to particular cases is achieved with unequal success. For a few of the simplest methods of representation there are no difficulties. For others one is equally well off using i nfinite series from the start. Additionally one can set

$$\begin{array}{r|l} \frac{y}{x} = \text{tang } v, & \frac{\mu}{\lambda} = \text{tang } w\,, \\ y = \beta \sin v, & \mu = \alpha \sin w\,, \\ x = \beta \cos v, & \lambda = \alpha \cos w. \end{array}$$

Then one obtains

$$\begin{array}{l|l} x + y\sqrt{-1} = \beta \varepsilon^{+v\sqrt{-1}}, & \mu\sqrt{-1} + \lambda = \alpha e^{+w\sqrt{-1}}\,, \\ x - y\sqrt{-1} = \beta \varepsilon^{-v\sqrt{-1}}, & \mu\sqrt{-1} - \lambda = -\alpha e^{-w\sqrt{-1}}\,; \end{array}$$

or using φ to denote the function

$$\beta \varepsilon^{+v\sqrt{-1}} = \varphi\,(\alpha e^{+w\sqrt{-1}})\,,$$
$$\beta \varepsilon^{-v\sqrt{-1}} = \varphi\,(-\alpha e^{-w\sqrt{-1}})\,.$$

Here it follows that one can, e.g., set

$$\beta \varepsilon^{+v\sqrt{-1}} = A + B\alpha^2 e^{2w\sqrt{-1}} + C\alpha^4 e^{4w\sqrt{-1}} + \text{etc.}\,,$$
$$\beta \varepsilon^{-v\sqrt{-1}} = A + B\alpha^2 e^{-2w\sqrt{-1}} + C\alpha^4 e^{-4w\sqrt{-1}} + \text{etc.}$$

from which

$$x = \beta \cos v = A + B\alpha^2 \cos 2w + C\alpha^4 \cos 4w + \text{etc.}\,,$$
$$y = \beta \sin v = \qquad B\alpha^2 \sin 2w + C\alpha^4 \sin 4w + \text{etc.}$$

can be found using known formulae. And here

$$w = \text{Arc tang}\left(\frac{\log \text{tang}\,(45^\circ + \frac{1}{2}p)}{\lambda}\right).$$

§ 74

Since one eventually relies on infinite series using this procedure, I return directly to the two differential equations (§ 70)

$$dy = M dp + m d\lambda\,,$$
$$dx = N dp + n d\lambda\,.$$

The circumstance, that here

$$+ M \cos p = n\,,$$
$$- N \cos p = m$$

that is, that both M and N must be multiplied by $\cos p$, guides us to the most proper form which can be assumed for the infinite series for x and y. Namely we take

$$\begin{aligned} y = A &\quad + B\lambda &\quad + C\lambda^2 &\quad + \text{etc.} \\ + A' \sin p &\quad + B' \lambda \sin p &\quad + C' \lambda^2 \sin p &\quad + \text{etc.} \\ + A'' \sin 2p &\quad + B'' \lambda \sin 2p &\quad + C'' \lambda^2 \sin 2p &\quad + \text{etc.} \\ + A''' \sin 3p &\quad + B''' \lambda \sin 3p &\quad + C''' \lambda^2 \sin 3p &\quad + \text{etc.} \\ + \text{etc.} \end{aligned}$$

and

$$\begin{aligned} x = a &\quad + b\lambda &\quad + c\lambda^2 &\quad + \text{etc.} \\ + a' \cos p &\quad + b' \lambda \cos p &\quad + c' \lambda^2 \cos p &\quad + \text{etc.} \\ + a'' \cos 2p &\quad + b'' \lambda \cos 2p &\quad + c'' \lambda^2 \cos 2p &\quad + \text{etc.} \\ + a''' \cos 3p &\quad + b''' \lambda \cos 3p &\quad + c''' \lambda^2 \cos 3p &\quad + \text{etc.} \\ + \text{etc.} \end{aligned}$$

§ 75

If y as well as x are now differentiated, one can find M, m, N, n and the equations

$$+ M \cos p = n\,,$$
$$- N \cos p = m$$

can be displayed explicitly. Inspecting the first of these, performing the indicated operations and accompanying simplification, yields

$$
\begin{aligned}
& M\cos p \\
&= \tfrac{1}{2}A' && + \tfrac{1}{2}B'\lambda && + \text{etc.} \\
&+ \tfrac{1}{2}(2A'')\cos p && + \tfrac{1}{2}(2B'')\lambda\cos p && + \text{etc.} \\
&+ \tfrac{1}{2}(A' + 3A''')\cos 2p && + \tfrac{1}{2}(B' + 3B''')\lambda\cos 2p && + \text{etc.} \\
&+ \tfrac{1}{2}(2A'' + 4A'''')\cos 3p && + \tfrac{1}{2}(2B'' + 4B'''')\lambda\cos 3p && + \text{etc.} \\
&+ \tfrac{1}{2}(3A''' + 5A^{V})\cos 4p && + \tfrac{1}{2}(3B''' + 5B^{V})\lambda\cos 4p && + \text{etc.} \\
&\quad \text{etc.}
\end{aligned}
$$

and

$$
\begin{aligned}
n = b & \qquad + 2c\lambda && + 3d\lambda^2 + \text{etc.} \\
+ b' & \cos p + 2c' \; \lambda\cos p && + 3d' \; \lambda^2\cos p \\
+ b'' & \cos 2p + 2c'' \; \lambda\cos 2p && + 3d'' \; \lambda^2\cos 2p \\
+ b''' & \cos 3p + 2c''' \; \lambda\cos 3p && + 3d''' \; \lambda^2\cos 3p
\end{aligned}
$$

§ 76

Here the coefficients can be compared term by term; one finds

$b = \tfrac{1}{2}A'$	$c = \tfrac{1}{4}B'$	$d = \tfrac{1}{6}C'$
$b' = \tfrac{1}{2}(2A'')$	$c' = \tfrac{1}{4}(2B'')$	$d' = \tfrac{1}{6}(2C'')$
$b'' = \tfrac{1}{2}(A' + 3A''')$	$c'' = \tfrac{1}{4}(B' + 3B''')$	$d'' = \tfrac{1}{6}(C' + 3C''')$
$b''' = \tfrac{1}{2}(2A'' + 4A'''')$	$c''' = \tfrac{1}{4}(2B'' + 4B'''')$	$d''' = \tfrac{1}{6}(2C'' + 4C'''')$
$b'''' = \tfrac{1}{2}(3A''' + 5A^{V})$	$c'''' = \tfrac{1}{4}(3B''' + 5B^{V})$	$d'''' = \tfrac{1}{6}(3C''' + 5C^{V})$
etc.	etc.	etc.

§ 77

In an entirely equivalent manner, using the equation $-N\cos p = m$, one finds the values

$B = 0$	$C = 0$	$D = 0$
$B' = \frac{1}{2}(2a'')$	$C' = \frac{1}{4}(2b'')$	$D' = \frac{1}{6}(2c'')$
$B'' = \frac{1}{2}(a' + 3a''')$	$C'' = \frac{1}{4}(b' + 3b''')$	$D'' = \frac{1}{6}(c' + 3c''')$
$B''' = \frac{1}{2}(2a'' + 4a'''')$	$C''' = \frac{1}{4}(2b'' + 4b'''')$	$D''' = \frac{1}{6}(2c'' + 4c'''')$
$B'''' = \frac{1}{2}(3a''' + 5a^{V})$	$C'''' = \frac{1}{4}(3b''' + 5b^{V})$	$D'''' = \frac{1}{6}(3c''' + 5c^{V})$
etc.	etc.	etc.

§ 78

The rule for the propagation in these expressions is very simple. One also easily finds that they depend on each other in alternating fashion, namely

the b's on the A's
the c's on the B's, therefore on the a's
the d's on the C's, therefore on the A's ,
and so on ,

the B's on the a's
the C's on the b's, therefore on the A's
the D's on the c's, therefore on the a's
and so on,

and it is seen that all the coefficients depend on a, A.

§ 79

One finds as a consequence

$$
\begin{array}{ll|ll}
b & = \frac{1}{2}(* + A') & B & = 0 \\
b' & = \frac{1}{2}(* + 2A'') & B' & = \frac{1}{2}(* + 2a'') \\
b'' & = \frac{1}{2}(A' + 3A''') & B'' & = \frac{1}{2}(a' + 3a''') \\
b''' & = \frac{1}{2}(2A'' + 4A'''') & B''' & = \frac{1}{2}(2a'' + 4a'''') \\
b'''' & = \frac{1}{2}(3A''' + 5A^{V}) & B'''' & = \frac{1}{2}(3a''' + 5a^{V}) \\
 & \text{etc.} & & \text{etc.}
\end{array}
$$

and, more generally

$$
b^k = \tfrac{1}{2}\,[(k-1)\,A^{k-1} + (k+1)\,A^{k+1}]\,,
$$
$$
B^k = \tfrac{1}{2}\,[(k-1)\,a^{k-1} + (k+1)\,a^{k+1}]\,.
$$

Furthermore

$$
\begin{array}{ll|ll}
c & = \frac{1}{2\cdot 4}(* \quad * \quad + 2a'') & C & = 0 \\
c' & = \frac{1}{2\cdot 4}(* \quad + 2a' \quad + 6a''') & C' & = \frac{1}{2\cdot 4}(* \quad + 2A' \quad + 6A''') \\
c'' & = \frac{1}{2\cdot 4}(* \quad + 8a'' \quad + 12a'''') & C'' & = \frac{1}{2\cdot 4}(* \quad + 8A'' \quad + 12A'''') \\
c''' & = \frac{1}{2\cdot 4}(2a' \quad + 18a''' + 20a^{V}) & C''' & = \frac{1}{2\cdot 4}(2A' \quad + 18A''' + 20A^{V}) \\
c'''' & = \frac{1}{2\cdot 4}(6a'' \quad + 32a'''' + 30a^{VI}) & C'''' & = \frac{1}{2\cdot 4}(6A'' \quad + 32A'''' + 30A^{VI}) \\
c^{V} & = \frac{1}{2\cdot 4}(12a''' + 50a^{V} + 42a^{VII}) & C^{V} & = \frac{1}{2\cdot 4}(12A''' + 50A^{V} + 42A^{VII}) \\
c^{VI} & = \frac{1}{2\cdot 4}(20a'''' + 72a^{VI} + 56a^{VIII}) & C^{VI} & = \frac{1}{2\cdot 4}(20A'''' + 72A^{VI} + 56A^{VIII}) \\
 & \text{etc.} & & \text{etc.}
\end{array}
$$

And, more generally

$$c^k = \tfrac{1}{4}\ [(k-1)\,B^{k-1} + (k+1)\,B^{k+1}]$$

$$= \frac{1}{2\cdot 4}\begin{bmatrix}(k-1)\cdot(k-2)\,a^{k-2} + 2kka^k \\ + (k+1)\cdot(k+2)\,a^{k+2}\end{bmatrix},$$

$$C^k = \tfrac{1}{4}\ [(k-1)\,b^{k-1} + (k+1)\,b^{k+1}]$$

$$= \frac{1}{2\cdot 4}\begin{bmatrix}(k-1)\cdot(k-2)\,A^{k-2} + 2kkA^k \\ + (k+1)\cdot(k+2)\,A^{k+2}\end{bmatrix}.$$

Furthermore

$$\begin{aligned}
d &= \frac{1}{2\cdot4\cdot6}(\,* \quad * \quad + 2A' \quad + 6A''') \\
d' &= \frac{1}{2\cdot4\cdot6}(\,* \quad * \quad + 16A'' \quad + 24A'''') \\
d'' &= \frac{1}{2\cdot4\cdot6}(\,* \quad + 8A' \quad + 60A''' \quad + 60A^{V}) \\
d''' &= \frac{1}{2\cdot4\cdot6}(\,* \quad + 40A'' \quad + 152A'''' \quad + 120A^{VI}) \\
d'''' &= \frac{1}{2\cdot4\cdot6}(\,6A' \quad + 114A''' \quad + 310A^{V} \quad + 210A^{VII}) \\
d^{V} &= \frac{1}{2\cdot4\cdot6}(\,24A'' + 248A'''' + 552A^{VI} + 336A^{VIII})
\end{aligned}$$

etc.

And similarly for D', D'', etc., as functions of a', a'', a''', etc. In general

$$D^k = \tfrac{1}{6}[(k-1)\,c^{k-1} + (k+1)\,c^{k+1}]\,.$$

$$= \frac{1}{2\cdot4\cdot6}\left\{\begin{array}{l} (k-1)\cdot(k-2)\cdot(k-3)\cdot a^{k-3} \\ + \ (k-1)\cdot(3k^2 - 3k + 2)\,a^{k-1} \\ + \ (k+1)\cdot(3k^2 + 3k + 2)\,a^{k+1} \\ + \ (k+1)\cdot(k+2)\cdot(k+3)\cdot a^{k+3}\end{array}\right\}.$$

Furthermore

$$
\begin{array}{lllllll}
e & = \dfrac{1}{2\cdot 4\cdot 6\cdot 8} & (* & * & * & +16a'' & +24a'''') \\
e' & = \dfrac{1}{2\cdot 4\cdot 6\cdot 8} & (* & * & +16a' & +120a''' & +120a^{V}) \\
e'' & = \dfrac{1}{2\cdot 4\cdot 6\cdot 8} & (* & * & +136a'' & +480a'''' & +360a^{VI}) \\
e''' & = \dfrac{1}{2\cdot 4\cdot 6\cdot 8} & (* & +40a' & +576a''' & +1360a^{V} & +840a^{VII}) \\
e'''' & = \dfrac{1}{2\cdot 4\cdot 6\cdot 8} & (* & +240a'' & +1696a'''' & +3120a^{VI} & +1680a^{VIII})
\end{array}
$$

etc.

And similarly for E', E'', etc., as functions of A', A'', A''', etc. In general

$$
e^{k} = \tfrac{1}{8}\left[(k-1)\,D^{k-1} + (k+1)\,D^{k+1}\right]
$$

$$
= \frac{1}{2\cdot 4\cdot 6\cdot 8}\left\{
\begin{array}{l}
\quad (k-1)\cdot(k-2)\cdot(k-3)\cdot(k-4)\,a^{k-4} \\
+\ (k-1)\cdot(k-2)\cdot(4k^{2}-8k+8)\,a^{k-2} \\
+\ (k\cdot k\cdot(6k^{2}+10)\,a^{k} \\
+\ (k+1)\cdot(k+2)\cdot(4k^{2}+8k+8)\,a^{k+2} \\
+\ (k+1)\cdot(k+2)\cdot(k+3)\cdot(k+4)\cdot a^{k+4}
\end{array}
\right\}.
$$

By this procedure the coefficients that follow can also be determined. I will end with this observation, since it is sufficient, generally speaking, to have shown that the computation proceeds smoothly even with the most complicated series. In particular cases there are noticeable abbreviations, as we will now see.

VII. APPLICATION OF THE METHOD TO A SPECIAL CASE

§ 80

There are several terrestrial maps, including ones of entire parts of the earth, for which their makers, presumably in order to save time, used the following method. One constructs a straight line down the middle of the map, and divides this into as many equal pieces as the meridian that it is to represent has degrees of latitude. Through each degree perpendicular lines are drawn, and these represent the circles parallel to the equator. These are also divided into equal parts, to represent the degrees of longitude. The degrees of longitude become smaller the closer the parallel is to the pole; specifically, in proportion to the sine of the distance from the pole, so that the degrees are made smaller along these lines in the same proportion. The remaining mid-day circles can thereby be drawn. These are crooked lines, which cross the equator at right angles, but all other parallels at oblique angles, thereby giving the countries a more or less obliquely stretched position. The lines themselves are those designated sinusoids by Leibnitz, and differ from each other only in that the ordinates are increased proportionally.

§ 81

We will here preserve two characteristics of this design; first that the middle meridian shall be a straight line, equally subdivided, and then that the equator shall be a straight line, perpendicularly cutting the meridian. To these requirements we however also add that all meridians shall penetrate all parallels at right angles, and that the degrees of latitude everywhere are in correct proportion to the degrees of longitude. These latter two conditions have as a consequence that the objective is contained within the much more general version investigated previously (§ 65 on). The first two conditions on the other hand serve to determine the form of the unending series which must be invoked to assist in the solution. Since I originally began with this approach, I took for y and x such series as are expanded in powers of p and λ. Now let, in figure 12, CA be the equator, and CB the midmost meridian. CE is equal to p, and CL a function of λ. The angle at E must be a right angle and all parallels on either side of CB must be similar.

This required that y be expressed as only even powers of λ and odd powers of p, whereas x must be made up of odd powers of λ, and even powers of p, and in such a manner that when $p = 0$ then $y = 0$ and x is a function of λ only; furthermore, when $\lambda = 0$, then $y = p$ and $x = 0$ must hold.

§ 82

The calculation which I hereupon performed yielded

$$
\begin{aligned}
y = p & \\
& + \tfrac{1}{2}\lambda^2 p - \tfrac{4}{2}\cdot\frac{1}{1\cdot2\cdot3}\lambda^2 p^3 + \frac{16}{2}\cdot\frac{\lambda^2 p^5}{1\cdot2\cdot3\cdot4\cdot5} - \frac{64}{2}\cdot\frac{\lambda^2 p^7}{1\cdot2\cdot3\cdot4\cdot5\cdot6\cdot7} + \text{etc.} \\
& + \tfrac{5}{24}\lambda^4 p - \tfrac{7}{18}\lambda^4 p^3 + \tfrac{5}{18}\lambda^4 p^5 - \text{etc.} \\
& + \tfrac{61}{720}\lambda^6 p - \tfrac{331}{1080}\lambda^6 p^3 + \text{etc.} \\
& + \tfrac{277}{8064}\lambda^8 p - \text{etc.} \\
& + \text{etc.}
\end{aligned}
$$

And

$$
\begin{aligned}
x = {} & \lambda - \tfrac{1}{2}\lambda p^2 + \frac{1}{2\cdot3\cdot4}\lambda p^4 - \frac{1}{2\cdot3\cdot4\cdot5\cdot6} - \lambda p^6 + \text{etc.} \\
& + \tfrac{1}{6}\lambda^3 - \tfrac{5}{12}\lambda^3 p^2 + \tfrac{41}{144}\lambda^3 p^4 - \tfrac{73}{864}\lambda^3 p^6 + \text{etc.} \\
& + \tfrac{1}{24}\lambda^5 - \tfrac{61}{240}\lambda^5 p^2 + \tfrac{1141}{2880}\lambda^5 p^4 - \text{etc.} \\
& + \tfrac{61}{5040}\lambda^7 - \tfrac{277}{2016}\lambda^7 p^2 + \text{etc.} \\
& + \tfrac{277}{72576}\lambda^9 - \text{etc.} \\
& + \text{etc.}
\end{aligned}
$$

§ 83

But as I summed these series it appeared, initially only by induction, that

$$
\begin{aligned}
y = {} & p \\
& + \tfrac{1}{4}\lambda^2 \sin 2p \\
& + \frac{4 \sin 2p + 3 \sin 4p}{4\cdot 8\cdot 1\cdot 3} \cdot \lambda^4 \\
& + \frac{34 \sin 2p + 60 \sin 4p + 30 \sin 6p}{4\cdot 8\cdot 12\cdot 1\cdot 3\cdot 5} \cdot \lambda^6 \\
& + \frac{496 \sin 2p + 1512 \sin 4p + 1680 \sin 6p + 630 \sin 8p}{4\cdot 8\cdot 12\cdot 16\cdot 1\cdot 3\cdot 5\cdot 7} \cdot \lambda^8 \\
& + \frac{11056 \sin 2p + 50880 \sin 4p + 93240 \sin 6p + 75600 \sin 8p + 22680 \sin 10p}{4\cdot 8\cdot 12\cdot 16\cdot 20\cdot 1\cdot 3\cdot 5\cdot 7\cdot 9} \cdot \lambda^{10} \\
& + \text{etc.,}
\end{aligned}
$$

and

$$
\begin{aligned}
x = {} & \cos p \cdot \lambda \\
& + \frac{\cos p + \cos 3p}{4\cdot 1\cdot 3} \cdot \lambda^3 \\
& + \frac{4 \cos p + 10 \cos 3p + 6 \cos 5p}{4\cdot 8\cdot 1\cdot 3\cdot 5} \cdot \lambda^5 \\
& + \frac{34 \cos p + 154 \cos 3p + 210 \cos 5p + 90 \cos 7p}{4\cdot 8\cdot 12\cdot 1\cdot 3\cdot 5\cdot 7} \cdot \lambda^7 \\
& + \frac{496 \cos p + 3520 \cos 3p + 8064 \cos 5p + 7560 \cos 7p + 2520 \cos 9p}{4\cdot 8\cdot 12\cdot 16\cdot 1\cdot 3\cdot 5\cdot 7\cdot 9} \cdot \lambda^9 \\
& + \text{etc.}
\end{aligned}
$$

I then assumed series of this form with indeterminate coefficients and thus found the induction to be quite correct.

§ 84

This is not where I left the matter, but observed that these values for y and x can be summed from below, and it appeared that

$$y = p + \sin 2p \,\text{tang}^2 \tfrac{1}{2}\lambda + \tfrac{1}{2} \sin 4p \,\text{tang}^4 \tfrac{1}{2}\lambda + \tfrac{1}{3} \sin 6p \,\text{tang}^6 \tfrac{1}{2}\lambda + \tfrac{1}{4} \sin 8p \,\text{tang}^8 \tfrac{1}{2}\lambda + \text{etc.},$$

$$x = 2 \cos p \,\text{tang} \tfrac{1}{2}\lambda + \tfrac{2}{3} \cos 3p \,\text{tang}^3 \tfrac{1}{2}\lambda + \tfrac{2}{5} \cos 5p \,\text{tang}^5 \tfrac{1}{2}\lambda + \text{etc.}$$

§ 85

Finally these series can also be summed, and it was found that

$$y = p + \text{Arc tang} \frac{\sin 2p \,\text{tang}^2 \frac{1}{2} \lambda}{1 - \cos 2p \,\text{tang}^2 \frac{1}{2} \lambda},$$

$$x = \tfrac{1}{2} \log \frac{1 + 2 \,\text{tang} \frac{1}{2} \lambda \cos p + \text{tang}^2 \frac{1}{2} \lambda}{1 - 2 \,\text{tang} \frac{1}{2} \lambda \cos p + \text{tang}^2 \frac{1}{2} \lambda}$$

or, if one sets $p = 90^\circ - \varepsilon$

$$y = 90^\circ - \varepsilon + \text{Arc cotg} \left(\text{cotg}\, 2\varepsilon + \text{cotg}^2 \tfrac{1}{2} \lambda \,\text{cosec}\, 2\varepsilon\right),$$

$$x = \tfrac{1}{2} \log \left(\frac{1 + \sin \lambda \cdot \sin \varepsilon}{1 - \sin \lambda \cdot \sin \varepsilon} \right) = \tfrac{1}{2} \log \left(\frac{\text{cosec}\, \varepsilon + \sin \lambda}{\text{cosec}\, \varepsilon - \sin \lambda} \right),$$

or, more concisely

$$\cot y = \cos \lambda \,\text{tang}\, \varepsilon .$$

§ 86

From this it was subsequently found that

$$dy = \frac{\cos \lambda \cdot dp + \sin p \cos p \sin \lambda \cdot dp}{1 - \cos^2 p \sin^2 \lambda)},$$

$$dx = \frac{\cos p \cos \lambda \cdot d\lambda - \sin \lambda \sin p \cdot dp}{1 - \cos^2 p \sin^2 \lambda},$$

whence

$$+M = \frac{\cos\lambda}{1-\cos^2 p \sin^2\lambda} \qquad + m = \frac{\sin p \cos p \sin\lambda}{1-\cos^2 p \sin^2\lambda}$$

$$-N = \frac{\sin p \sin\lambda}{1-\cos^2 p \sin^2\lambda} \qquad + n = \frac{\cos\lambda \cos p}{1-\cos^2 p \sin^2\lambda},$$

where one easily sees that these expressions both satisfy the fundamental equations (§ 69):

$$+M\cos p = n, \; -N\cos p = m$$

and therefore also preserve the induction employed for the summation.

§ 87

From the formula

$$x = \tfrac{1}{2}\log\left(\frac{1+\sin\varepsilon\sin\lambda}{1-\sin\varepsilon\sin\lambda}\right)$$

it is apparent that x depends on ε as it does on λ so that one, for example, obtains the same x value for $\varepsilon = 40^{\circ}, \lambda = 60^{\circ}$, as for $\varepsilon = 60^{\circ}, \lambda = 40^{\circ}$.

§ 88

If one sets $\varepsilon = 90^{\circ}$, or $p = 0$, then

$$x = \tfrac{1}{2}\log\left(\frac{1+\sin\lambda}{1-\sin\lambda}\right) = \log \operatorname{tang}(45^{\circ} + \tfrac{1}{2}\lambda).$$

Consequently the degrees of the equator measured from C (of figure 12), increase exactly as do the degrees of latitude on Mercator's nautical charts. In fact the entire representation, as we shall see in the sequel, parallels that of Mercator.

§ 89

If one now sets $\lambda = 90^{\circ}$, then

$$x = \tfrac{1}{2}\log\frac{1+\sin\varepsilon}{1-\sin\varepsilon} = \log \operatorname{tang}(45^{\circ} + \tfrac{1}{2}\varepsilon)$$

and $y = 90^o$. Consequently the meridian situated at 90° from CB (figure 12), is parallel to the equator, and the degrees of equatorial distance, measured from the pole, increase thereon exactly as do the degrees of longitude along the equator measured from C, or as the degrees of latitude on Mercator's nautical charts. The consequence of this is that all of the parallel circles intersect all of the meridional circles at right angles, and are pulled up from E towards M, and become parallel to CB at 90° of longitude at D. Above D they are again curved. And since $DB > BE$ they are regular ovals, becoming more oblong the further they are from the pole.

§ 90

One finds additionally (figure 12) for every parallel circle the radius of the circle of curvature

$$\text{at } E = \text{tang}\ \varepsilon, \text{at } D = \sin \varepsilon\ ,$$

and for every meridional circle BML the radius of the circle of curvature at

$$L = \text{cosec}\ \lambda\ .$$

Finally, the meridional circles have an inflextion point at B. Their radius of curvature is here infinite, and this requires that they pass through the pole in a very straight direction. In general they are less distinguishable from the aforementioned sine curves, the smaller is λ.

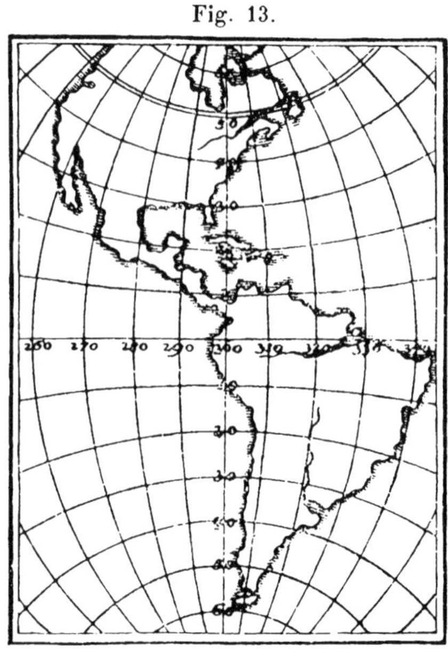

Fig. 13.

§ 91

In the accompanying table I have computed and presented the values of x and y by 10 degree increments of ε and λ, and have used this for a representation of America, as shown in the thirteenth figure.

To Accompany §91

For x

	$\lambda = 0$	10	20	30	40	50	60	70	80	90
$\varepsilon = 0$	0,00 000	0,00 000	0,00 000	0,00 000	0,00 000	0,00 000	0,00 000	0,00 000	0,00 000	0,00 000
10	0,00 000	0,03 016	0,05 946	0,08 704	0,11 209	0,13 382	0,15 153	0,16 465	0,17 271	0,17 543
20	0,00 000	0,05 946	0,11 752	0,17 271	0,22 363	0,26 830	0,30 531	0,33 323	0,35 051	0,35 638
30	0,00 000	0,08 704	0,17 271	0,25 541	0,33 697	0,40 360	0,46 360	0,50 987	0,53 928	0,54 931
40	0,00 000	0,11 209	0,22 363	0,33 697	0,43 944	0,53 923	0,62 805	0,69 946	0,74 644	0,76 291
50	0,00 000	0,13 382	0,26 830	0,40 360	0,53 923	0,67 283	0,79 889	0,90 733	0,98 310	1,01 068
60	0,00 000	0,15 153	0,30 531	0,46 360	0,62 805	0,79 889	0,97 296	1,13 817	1,26 892	1,31 694
70	0,00 000	0,16 465	0,33 323	0,50 987	0,69 946	0,90 733	1,13 817	1,38 939	1,62 549	1,73 542
80	0,00 000	0,17 271	0,35 051	0,53 928	0,74 644	0,98 310	1,26 892	1,62 549	2,08 925	2,43 624
90	0,00 000	0,17 543	0,35 638	0,54 931	0,76 291	1,01 068	1,31 694	1,73 542	2,43 624	infinit.

For y

	$\lambda = 0$	10	20	30	40	50	60	70	80	90
$\varepsilon = 0$	1,57 080	1,57 080	1,57 080	1,57 080	1,57 080	1,57 080	1,57 080	1,57 080	1,57 080	1,57 080
10	1,39 626	1,39 886	1,40 648	1,41 926	1,43 670	1,45 793	1,46 541	1,51 056	1,54 018	1,57 080
20	1,22 173	1,22 643	1,24 125	1,26 545	1,29 888	1,34 097	1,39 078	1,44 695	1,50 364	1,57 080
30	1,04 720	1,05 380	1,07 370	1,10 712	1,16 690	1,21 348	1,28 977	1,37 584	1,47 088	1,57 080
40	0,87 266	0,88 019	0,90 311	0,94 239	0,99 951	1,07 617	1,17 367	1,29 133	1,42 611	1,57 080
50	0,69 813	0,70 568	0,72 891	0,76 961	0,83 088	0,91 699	1,03 361	1,18 375	1,36 703	1,57 080
60	0,52 360	0,53 025	0,55 094	0,58 800	0,64 585	0,73 182	0,85 707	1,03 599	1,27 464	1,57 080
70	0,34 907	0,35 401	0,36 953	0,39 782	0,44 355	0,51 522	0,62 923	0,81 648	1,10 100	1,57 080
80	0,17 453	0,17 717	0,18 542	0,20 086	0,22 624	0,27 627	0,33 904	0,48 601	0,79 305	1,57 080
90	0,00 000	0,00 000	0,00 000	0,00 000	0,00 000	0,00 000	0,00 000	0,00 000	0,00 000	

The drawing appears most natural since the spacing of parallels along the 30th degree is even, and the degrees of longitude are quite uniform. They would have been very uneven if the map had had to extend further in longitude on either side. But fortunately America permits enclosure within 80 degrees of longitude, measured at the equator. Europe and Africa fare even better in this type of representation, since even fewer degrees of longitude are required. Finally Asia comes out satisfactorily if one takes the 90th meridian as the central one, and does not retain more than from 50 degrees to 135 degrees along the equator.

§ 92

This method of representation also coincides quite completely with that of Mercator. The difference is only that, in Mercator's case the equator is BC (figure 12) and DB, RM, CA are meridional circles. We earlier (§ 85) found

$$x = \tfrac{1}{2} \log \left(\frac{1 + \sin \lambda \sin \varepsilon}{1 - \sin \lambda \sin \varepsilon} \right)$$

Since, in general, ε represents the arc BM, and λ the angle MBR, then

$$\sin \lambda \sin \varepsilon = \sin \varphi \; ,$$

therefore φ is the arc of the great circle drawn perpendicular to BC through the point M. Then, inserting this value

$$MR = x = \tfrac{1}{2} \log \left(\frac{1 + \sin \varphi}{1 - \sin \varphi} \right) = \log \operatorname{tang} (45^\circ + \tfrac{1}{2} \varphi) \; .$$

Consequently, MR actually represents the arc φ of Mercator's representation.

VIII. REGULAR REPRESENTATIONS OF THE EARTH'S SURFACE

§ 93

Among the many portraits of the spherical surface, those that display the greatest regularity are those on which the meridians are represented by straight lines intersecting at the proper angle at the pole, and all equally subdivided away from the pole, so that the equator, and the latitudes parallel to it, are circles with a common center at the pole. The third figure demonstrates such a representation. The only exceptional characteristic there is that the radii of the parallels increase as the tangent of half the distance to the pole. But it is easily seen that they could increase according to some alternate arbitrary rule without destroying the regularity.

§ 94

It is immaterial whether the midpoint p is to represent the pole or some other location on the earth's surface. For any point can be regarded as the pole on a sphere. But if p is not the pole then all the lines receive different names. The meridional circles become vertical circles, the equator becomes the horizon, and the latitude circles become height circles; finally the degrees of the equator become azimuthal arcs or degrees of world regions. This follows directly from the definitions of the terms. One can also envision the situation as follows. Let, for example, (figure 3) p be Berlin, pA its meridian, then the angles apA, bpA, γpA show how many degrees the places a, c, b, γ lie to the west or east of the meridian, and ap, cp, bp, γp measured in degrees, give the separation of these places from Berlin in degrees. If instead of such arbitrary places, one takes the intersections of all meridians and parallels, then the angles apA, bpA and so on, and the degrees ap, bp, etc., can be found by known trigonometric calculations, and the meridians and parallels can therefore be drawn to completely preserve the regularity of the representation.

§ 95

For representations of hemispheres of the earth one usually takes the 90th and 270th degrees of longitude on the equator as the midpoints p, and the two poles therefore coincide with B and A and DpE becomes the equator. This results in a completely similar appearing drawing for the meridians and parallels in the four quarters DpB, EpB, EpA, DpA.

§ 96

Let, for example, using the fourteenth figure, ACE be the equator; P, p the poles; M the intersection of the meridian $PMLp$ and latitude circle MB. Now since the representation is to be established so that all straight lines through the midpoint C represent greatest circles of the sphere, and are to be divided in a consistent manner, then CM is an arc of such a greatest circle, which gives the distance of M from C in degrees, and the angle ACM determines the position of M relative to the equator.

Fig. 14.

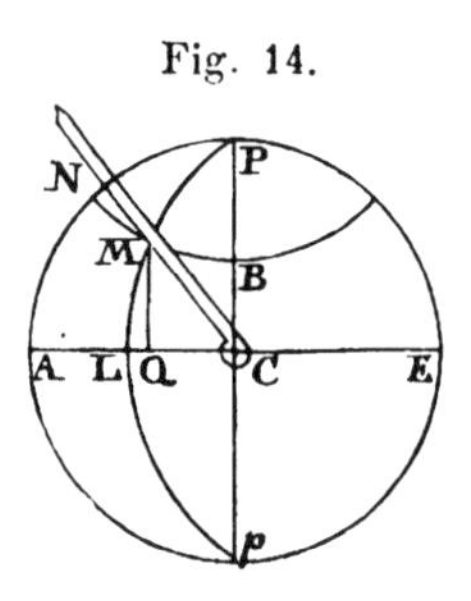

§ 97

Now set the difference in

$$\begin{aligned} \text{longitude } CL &= \lambda; \\ \text{the latitude } LM \text{ or } CB &= p; \\ \text{the angle } ACM &= w; \\ \text{the separation } CM &= k; \end{aligned}$$

then

$$\cos k = \cos \lambda \cos p$$
$$\operatorname{tang} \omega = \operatorname{tang} p : \sin \lambda$$

The accompanying tables have been constructed according to these formulae, and give k and ω for all arcs p, λ by 5-degree increments. The use is as follows.

To Accompany §97

For ω

	$p=0$	5	10	15	20	25	30	35	40	45	50	55	60	65	70	75	80	85	90
$\lambda=0$		90°0′	90°0′	90°0′	90°0′	90°0′	90°0′	90°0′	90°0′	90°0′	90°0′	90°0′	90°0′	90°0′	90°0′	90°0′	90°0′	90°0′	90°0′
5	0.0	15. 7	63.42	71.59	76.32	79.25	81.25	82.54	84. 4	85. 1	85.49	86.30	87. 7	87.47	88.11	88.40	89. 7	89.35	90.0
10	0.0	26.44	45.26	57. 3	64.30	69.35	73.16	76. 4	78.18	80. 9	81.43	83. 4	84.16	85.22	86.23	87.20	88.15	89. 8	90.0
15	0.0	18.40	34.16	46. 0	54.35	60.58	65.51	69.43	72.46	75.29	77.45	79.44	81.30	83. 7	84.37	86. 2	87.23	88.42	90.0
20	0.0	14.21	27.16	38. 5	46.47	53.45	59.21	63.58	67.49	71. 7	73.59	76.32	78.50	80.56	82.54	84.46	86.33	88.17	90.0
25	0.0	10.56	22.39	32.22	40.44	47.49	53.48	58.53	63.16	67. 5	70.29	73.31	76.17	78.51	81.15	83.32	85.44	87.55	90.0
30	0.0	9.54	19.26	28.11	36. 3	43. 0	49. 6	54.28	59.13	63.26	67.14	70.42	73.54	76.53	79.41	82.22	84.58	87.30	90.0
35	0.0	8.40	17. 6	25. 2	32.24	39. 7	45.11	50.41	55.39	60.10	64.18	68. 7	71.41	75. 2	78.12	81.16	84.14	87. 8	90.0
40	0.0	7.45	15.20	22.38	29.31	35.58	41.56	47.27	52.33	57.16	61.39	65.46	69.38	73.39	76.50	80.14	83.32	86.47	90.0
45	0.0	7. 3	14. 0	20.45	27.14	33.24	39.14	44.43	49.53	54.44	59.19	63.39	67.48	71.45	75.34	79.16	82.53	86.28	90.0
50	0.0	6.31	12.58	19.17	25.25	31.20	37. 0	42.25	47.36	52.33	57.16	61.48	66. 9	70.21	74.25	78.24	82.18	86.10	90.0
55	0.0	6. 6	12. 9	18. 7	23.58	29.39	35.11	40.32	45.41	50.41	55.30	60.10	64.41	69. 6	73.45	77.37	81.47	85.54	90.0
60	0.0	5.46	11.31	17.12	22.48	28.18	33.42	38.57	44. 6	49. 6	54. 0	58.46	63.26	68. 1	72.30	76.56	81.19	85.40	90.0
65	0.0	5.31	11. 1	16.28	21.53	27.13	32.30	37.41	42.48	47.49	52.45	57.36	62.23	67. 5	71.45	76.21	80.55	85.28	90.0
70	0.0	5.19	10.38	15.55	21.10	26.24	31.34	36.41	41.46	46.47	51.45	56.39	61.31	66.20	71. 7	75.52	80.36	85.18	90.0
75	0.0	5.11	10.21	15.31	20.39	25.46	30.52	35.56	40.47	46. 0	50.59	55.56	60.51	65.45	70.38	75.29	80.20	85.10	90.0
80	0.0	5. 5	10. 9	15.13	20.17	25.20	30.23	35.25	40.26	45.26	50.26	55.25	60.23	65.20	70.17	75.13	80. 9	85. 5	90.0
85	0.0	5. 1	10. 2	15. 4	20. 4	25. 5	30. 6	35. 6	40. 6	45. 7	50. 6	55. 6	60. 6	65. 5	70. 4	75. 3	80. 2	85. 1	90.0
90	0.0	5. 0	10. 0	15. 0	20. 0	25. 0	30. 0	35. 0	40. 0	45. 0	50. 0	55. 0	60. 0	65. 0	70. 0	75. 0	80. 0	85. 0	90.0

To Accompany §97

For k

	$p=0$	5	10	15	20	25	30	35	40	45	50	55	60	65	70	75	80	85	90
$\lambda=0$	0°0′	5°0′	10°0′	15°0′	20°0′	25°0′	30°0′	35°0′	40°0′	45°0′	50°0′	55°0′	60°0′	65°0′	70°0′	75°0′	80°0′	85°0′	90°0
5	5.0	7. 4	11.10	15.48	20.35	25.28	30.23	35.19	40.16	45.13	50.11	55. 9	60. 8	65. 9	70. 5	75. 3	80. 2	85. 1	90.0
10	10.0	11.10	14. 6	17.58	22.16	26.48	31.29	36.13	41. 2	45.44	50.44	55.36	60.30	65.24	70.19	75.14	80. 9	85. 5	90.0
15	15.0	15.48	17.58	21. 6	24.49	28.54	33.14	37.42	42.16	46.55	51.37	56.21	61. 7	65.55	70.43	75.31	80.21	85.10	90.0
20	20.0	20.35	22.16	24.49	27.59	31.37	35.32	39.40	43.57	48.22	52.50	57.53	61.58	66.36	71.15	75.55	80.37	85.18	90.0
25	25.0	25.28	26.48	28.54	31.37	34.47	38.20	42. 4	46. 2	50. 2	54.22	58.41	63. 3	67.26	71.57	76.26	80.57	85.29	90.0
30	30.0	30.23	31.29	33.14	35.32	38.20	41.25	44.49	48.26	52.14	56.10	60.13	64.20	68.32	72.46	77. 3	81.21	85.40	90.0
35	35.0	35.19	36.13	37.42	39.40	42. 4	44.49	47.51	51. 8	54.36	58.13	61.58	65.46	69.45	73.44	77.46	81.49	85.54	90.0
40	40.0	40.16	41. 2	42.16	43.57	46. 2	48.26	51. 8	54. 4	57.12	60.30	63.56	67.29	71. 7	74.49	78.34	82.21	86.10	90.0
45	45.0	45.13	45.44	46.55	48.22	50. 2	52.14	54.36	57.12	60. 0	62.58	66. 4	69.18	72.37	76. 0	79.27	82.57	86.28	90.0
50	50.0	50.11	50.44	51.37	52.50	54.22	56.10	58.13	60.30	62.58	65.36	68.22	71.15	74.14	77.18	80.25	83.35	86.47	90.0
55	55.0	55. 9	55.36	56.21	57.23	58.41	60.13	61.58	63.56	66. 4	68.22	70.48	73.20	75.58	78.41	81.28	84.17	87. 8	90.0
60	60.0	60. 8	60.30	61. 7	61.58	63. 3	64.20	65.46	67.29	69.18	71.15	73.20	75.31	77.48	80. 9	82.34	85. 1	87.30	90.0
65	65.0	65. 9	65.24	65.55	66.36	67.29	68.32	69.45	71. 7	72.37	74.14	75.58	77.48	79.43	81.41	83.45	85.47	87.53	90.0
70	70.0	70. 5	70.19	70.43	71.15	71.57	72.46	73.44	74.49	76. 0	77.18	78.41	80. 9	81.41	83.17	84.55	86.36	88.17	90.0
75	75.0	75. 3	75.14	75.31	75.55	76.26	77. 3	77.46	78.34	79.27	80.25	81.28	82.34	83.45	84.55	86.10	87.26	88.42	90.0
80	80.0	80. 2	80. 9	80.21	80.37	80.57	81.21	81.49	82.21	82.57	83.35	84.17	85. 1	85.47	86.36	87.26	88.16	89. 8	90.0
85	85.0	85. 2	85. 5	85.10	85.18	85.29	85.40	85.29	86.10	86.28	86.47	87. 8	87.30	87.53	88.17	88.42	89. 8	89.34	90.0
90	90.0	90. 0	90. 0	90. 0	90. 0	90. 0	90. 0	90. 0	90. 0	90. 0	90. 0	90. 0	90. 0	90. 0	90. 0	90. 0	90. 0	90. 0	90.0

§ 98

Divide the circle $APEp$ in degrees. Then one assumes a convenient rule by which to subdivide into degrees the lines passing through C (that, as mentioned, are representations of greatest circles). These subdivisions are actually marked on a ruler, which is permitted to rotate about C, so that degrees from C can be counted. Having done this, the ruler is turned so that for every value of p and λ and the arc $AN = \omega$, the corresponding value of k is sought on the subdivision of CN, and the desired intersection of p and λ, is drawn through the point M. For example, if $\lambda = 50^{\circ}$ and $p = 30^{\circ}$ one finds in the table that

$$\omega = 37^{\circ}\,0', \; k = 56^{\circ}\,10'$$

and therefore

$$AN = 37^{\circ}\,0'$$

for M on CN use $56^{\circ}\,10'$.

§ 99

It now only depends on what rule is to be used for the specification of degrees on CN. If one takes

$$CM = \sin k\,,$$

so that all degrees are inserted as the sine of the distance from C toward N, then one obtains the orthographic projection. One obtains the stereographic projection if one makes

$$CM = \text{tang}\,\tfrac{1}{2}\,k\,.$$

If instead one takes $CM = \text{tang}\,k$ then the central projection is obtained. Since these representations are known I have taken

$$CM = k$$

as a variant, and therefore have divided CN into 90 equal pieces, the same as the number of degrees available. The representation which flows from this choice can be seen in the fifteenth figure. It has the advantage that the spacings along the meridians and parallels do not differ greatly, and that these curves do not intersect at very oblique angles. The intersection angles along AP, CP, CA are all 90 degrees and the spacings on CA as well as on CP are equal, as is the subdivision of AP. This method of representation has only the property that distances of all places from C can be measured with a uniformly subdivided scale. On the other hand, as we shall see subsequently, if one makes

Fig. 15.

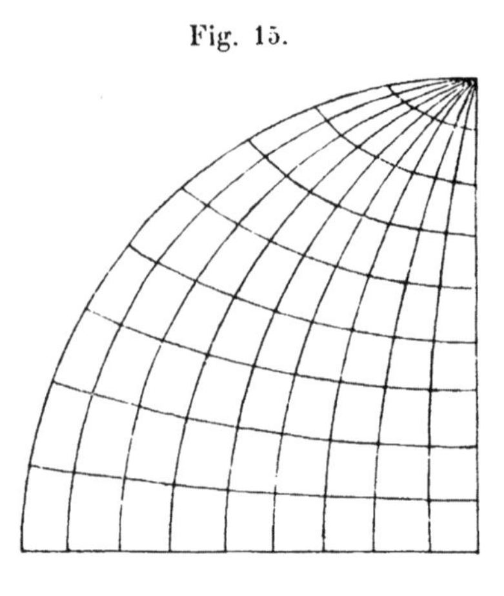

$$CM = \sin \tfrac{1}{2} k$$

and thus transcribes all degrees according to the sine of their halves from C toward N, then the distinct specific condition is obtained that all countries acquire a space in the drawing proportional to their true size.

IX. REPRESENTATIONS OF THE EARTH'S SURFACE CONSIDERING THE SIZES OF COUNTRIES

§ 100

It is not possible to consider the sizes of countries in any of the representations mentioned thus far, since they are devoted to the satisfaction of other considerations. On the stereographic, and even more so on the central projection, the degrees become larger away from the center, and therefore countries lying further from the center of the map appear much larger than they actually are. On Mercator's nautical charts those portions lying toward the poles become endlessly large. Contrarily, on the orthographic projection those portions lying further from the middle of the map become smaller, and the countries around the edge become endlessly small. Therefore if the question is one of representing the earth's surface so that all countries proportionately preserve their size, then the representation must be especially designed to do this.

§ 101

This can be achieved in very many various ways. The general solution to the question however is of no less difficulty and length then that posed earlier (§ 65). But we can present several of the simpler cases, which, because they are well specified, present fewer difficulties. The first is therefore to be that in which the meridional circles are straight parallel lines, which intersect the equator, and its parallels, which are also straight, at right angles. This case is very easy. The

Fig. 16.

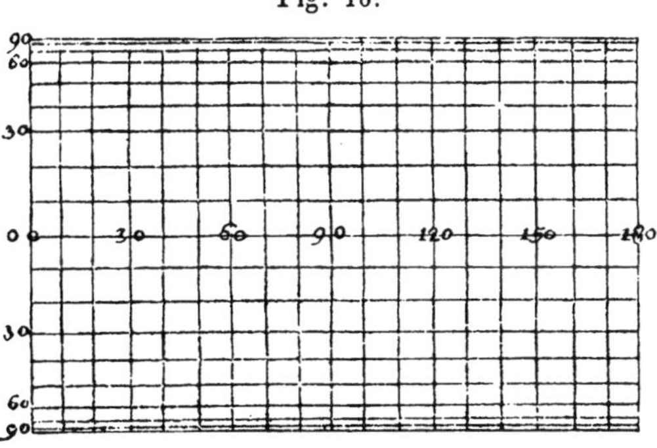

equator is divided into 360 equal parts, one for each degree, and the meridians are now straight lines cutting the equator at right angles, and the parallels are inserted where their sines fall above and below the equator. Figure 16 represents half the world in this fashion. The entire procedure rests on the fact that the zones of the earth from the equator to the poles increase their spatial content as the sine of the latitude. This requires that the degrees of latitude became noticeable smaller toward the pole, and endlessly small at the pole. However, since the first 30 degrees of latitude are not very different, a map of Africa, or of other countries lying near the equator, comes off well (see figure 17).

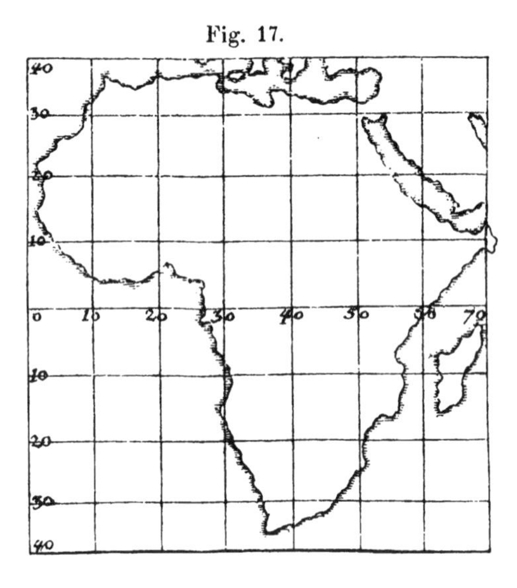

Fig. 17.

§ 102

On the other hand, for countries such as America, for example, which have their greatest dimension in a north-south direction, it is better to turn the method of representation around, in such a way that it is not the equator but the middle meridional circle which is represented by a straight line equally divided, and the equator is then divided according to the sine of the longitude. The 18th figure shows Asia represented in this manner.

To Accompany § 102

The Arcs *A R*

	$\lambda = 10$	20	30	40	50	60	70	80	90
$p = 10°0′$	10° 9′	10°38′	11°31′	12°58′	15°20′	19°26′	27°16′	45°26′	90°0′
20.0	20.17	21.10	22.48	25.25	29.31	36. 3	46.47	64.30	90.0
30.0	30.23	31.34	33.42	37. 0	41.56	49. 6	59.21	73.16	90.0
40.0	40.26	41.46	44. 6	47.36	52.33	59.13	67.49	78.18	90.0
50.0	50.26	51.45	54. 0	57.16	61.39	67.14	73.59	81.43	90.0
60.0	60.23	61.31	63.26	66. 9	69.38	73.54	78.50	84.16	90.0
70.0	70.17	71. 7	72.30	74.25	76.50	79.41	82.54	86.23	90.0
80.0	80. 9	80.36	81.19	82.18	83.32	84.58	86.33	88.15	90.0

The Arcs *A Q*

	$p = 80$	70	60	50	40	30	20	10	0
$\lambda = 10°$	1°44′	3°24′	4°59′	6°25′	7°39′	8°39′	9°23′	9°51′	10°0′
20	3.24	6.43	9.51	12.42	15.11	17.14	18.45	19.41	20.0
30	4.59	9.51	14.29	18.45	22.31	25.40	28. 2	29.30	30.0
40	6.25	12.42	18.45	24.24	29.30	33.50	37.10	39.16	40.0
50	7.39	15.11	22.31	29.30	35.56	41.34	46. 3	48.58	50.0
60	8.39	17.14	25.40	33.50	41.34	48.35	54.28	58.31	60.0
70	9.23	18.45	28. 2	37.10	46. 3	54.28	62. 1	67.44	70.0
80	9.51	19.41	29.30	39.16	48.58	58.31	67.44	75.54	80.0
90	10. 0	20. 0	30. 0	40. 0	50. 0	60. 0	70. 0	80. 0	90.0

AP is the equally divided meridian and goes through the middle of the map. DAB is the equator, subdivided as the sine of the longitude from A to B and to D. Each intersection point M is easily found. For the ordinate MR is the sine of the longitude or of the angle RPM, and PR is an arc whose tangent is equal to the product of the tangent of the equatorial height PM and the cosine of the angle RPM. Then the ordinate MR represents a greatest circle, which cuts the meridian PA perpendicularly at R. If one constructs MQ perpendicular to AB, then Q indicates the degrees where the sine of $AQ = RM$ on the segment AQ. I have presented, in the accompanying table, all 10 by 10 degree intersections, the two entries AR and AQ in degrees, for every point M, where p is the polar distance and λ is the longitude measured from A. It was not necessary to perform a special computation for this table, since it could be extracted from the earlier one (§ 97) requiring only changes in the headings.

Fig. 18.

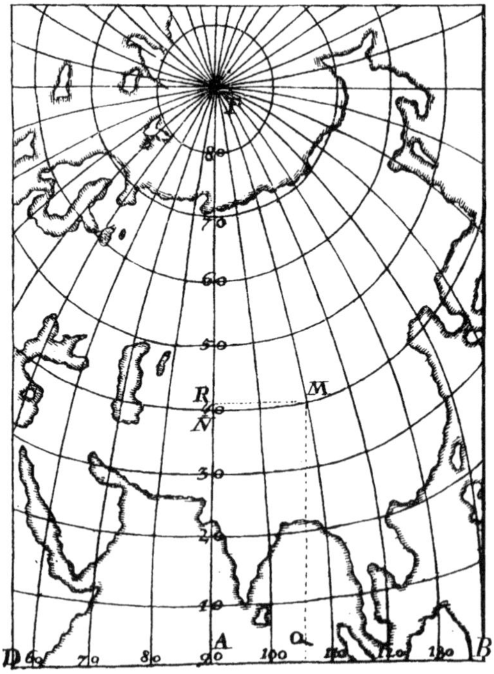

§ 103

I will now pass on to a different method of representation in which the meridians are straight lines intersecting at the proper angle at the pole. The problem is to subdivide these so that all countries preserve their size relationships.

§ 104

In the 8th figure (§ 48) let PN, pv be two infinitely close meridians, and let two infinitely close parallels $M\mu$, Nv be drawn with the pole P as their center. For the first of these let ε be the equatorial height, for the other let it be $\varepsilon + d\varepsilon$. The angle NPv is $d\lambda$, and one takes $PM = x$, $MN = dx$. Hence it follows that

$$M\mu = x d\lambda\,,$$

and thus the contents of

$$MNv\mu = x dx \cdot d\lambda\,.$$

These contents must equal those of the region they represent on the sphere, thus ,

$$x dx \cdot d\lambda = \sin\varepsilon \cdot d\varepsilon \cdot d\lambda$$

must hold. Consequently

$$x\,dx = \sin\varepsilon \cdot d\varepsilon = -\,d\cos\varepsilon\,,$$
$$\tfrac{1}{2}\,xx = \text{Const.} - \cos\varepsilon\,.$$

Now $x = 0$ should hold when $\varepsilon = 0$. Consequently ,

$$\tfrac{1}{2}\,xx = 1 - \cos\varepsilon = 2\sin^2\tfrac{1}{2}\,\varepsilon\,,$$
$$x = 2\sin\tfrac{1}{2}\,\varepsilon\,.$$

This then is the subdivision of the meridional circles, cited at the conclusion of § 99. Since we there already indicated the application, we can present such an application in the 19th figure, where two hemispheres of the earth are shown, and the poles are not in the center but at the top and bottom edges of the circle. The meridians as well as the parallels are curved lines, which can be constructed with two circles, since their ordinates are the differences of the ordinates of two circles.

Fig. 19.

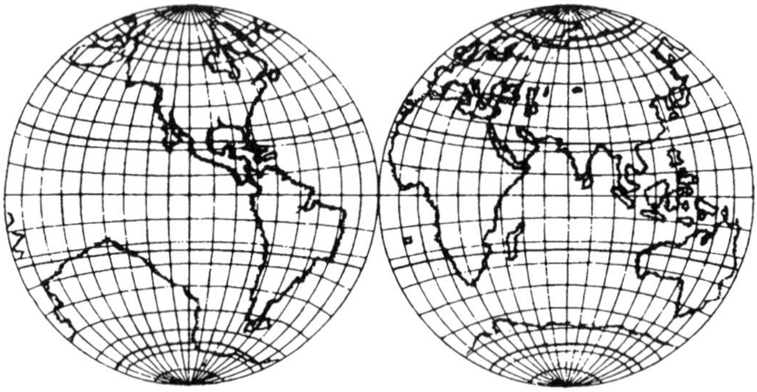

§ 105

In the 14th figure let everything have the same meaning as in § 96 and § 97. Then

$$CM = 2 \sin \tfrac{1}{2} k \,.$$

Draw the ordinate MQ, and set

$$QC = x$$
$$MQ = y$$

then (by § 97)

$$y : x = \text{tang}\, \omega = \text{tang}\, p : \sin \lambda \,.$$

Furthermore

$$x^2 + y^2 = CM^2 = 4 \sin^2 \tfrac{1}{2} k = 2\,(1 - \cos k) \,.$$

Now since (§ cited)

$$\cos k = \cos \lambda \cos p ,$$

so that

$$x^2 + y^2 = 2 - 2 \cos \lambda \cos p .$$

Now if either λ or p are eliminated from these equations

$$y \sin \lambda = x \operatorname{tang} p ,$$
$$x^2 + y^2 = 2 - 2 \cos \lambda \cos p$$

one obtains, in the first instance, an equation relating x, y, p, and this determines the curved line for each parallel whose latitude $= p$. In the other instance an equation relating x, y, λ is obtained, and this specifies the curved line for every meridian separated from C by λ degrees.

§ 106

The equation for the first case, the latitude curves, is

$$x^2 + y^2 = 2 - 2 \cos p \sqrt{1 - \frac{xx}{yy} \cdot \operatorname{tang}^2 p} .$$

From this one finds, after several simplifications,

$$y = \sqrt{1 + \sin p - \tfrac{1}{4}x^2} \pm \sqrt{1 - \sin p - \tfrac{1}{4}x^2}$$

or, if

$$p = 90^\circ - \varepsilon$$

is used

$$y = \sqrt{2 + \cos^2 \tfrac{1}{2}\varepsilon - \tfrac{1}{4}x^2} \pm \sqrt{2 \sin^2 \tfrac{1}{2}\varepsilon - \tfrac{1}{4}x^2} .$$

§ 107

In the other situation, the meridional case, the equation is

$$x = \sqrt{1 + \sin \lambda - \tfrac{1}{4}y^2 (1 + \sin \lambda)^2} \pm \sqrt{1 - \sin \lambda - \tfrac{1}{4}y^2 (1 - \sin \lambda)^2}$$

or, with

$$\lambda = 90^{\circ} - L$$

$$x = \cos^2 \tfrac{1}{2} L \cdot \sqrt{2 \sec^2 \tfrac{1}{2} L - y^2} \pm \sin^2 \tfrac{1}{2} L \cdot \sqrt{2 \operatorname{cosec}^2 \tfrac{1}{2} L - y^2} \; .$$

§ 108

In the current example we have thus far assumed that the meridians intersect at the correct angles at the pole. This condition can be modified to take the angles to be m-times larger or smaller than the true ones. Thus the equation (§ 104)

$$x\,dx \cdot d\lambda = \sin \varepsilon \, d\varepsilon \cdot d\lambda$$

is now changed to

$$m x\,dx \cdot d\lambda = \sin \varepsilon \, d\varepsilon \cdot d\lambda$$

which yields

$$m x\,dx = \sin \varepsilon \cdot d\varepsilon = - \, d \cos \varepsilon \, ,$$

therefore

$$\tfrac{1}{2}\, m x x = \text{Const.} - \cos \varepsilon$$

and, if the constant is specified as before,

$$\tfrac{1}{2}\, x x = \frac{1}{m} (1 - \cos \varepsilon) = \frac{2}{m} \sin^2 \tfrac{1}{2} \varepsilon \, ,$$

must yield

$$x = 2 \sin \tfrac{1}{2} \varepsilon \sqrt{\frac{1}{m}} \; .$$

§ 109

Since here one has the option of choosing m arbitrarily, it can be done as in the earlier case (§ 56, 57) to construct cones with $m = \frac{3}{4}$ or $= \frac{4}{5}$ or $= \frac{5}{6}$ and so on.

§ 110

But if one wishes to use the formula to draw individual countries, Europe for example, then m can be determined by requiring that the middle degree of latitude has the proper proportion with respect to the middle degree of longitude, or

$$mxd\lambda : dx = \sin\varepsilon' \cdot d\lambda : d\varepsilon' ,$$
$$mxd\lambda \cdot d\varepsilon' = \sin\varepsilon' \cdot dx \cdot d\lambda .$$

Now since

$$x = 2\sin\tfrac{1}{2}\varepsilon' : \sqrt{m} ,$$
$$dx = d\varepsilon' \cos\tfrac{1}{2}\varepsilon' : \sqrt{m} ,$$

from which it is seen that

$$m = \cos^2\tfrac{1}{2}\varepsilon' = \frac{1 + \cos\varepsilon'}{2} .$$

If one takes, for example as earlier (§ 52) for Europe

$$\cos\varepsilon' = \tfrac{3}{4} ,$$

then $m = \frac{7}{8}$ and this is the ratio by which the degrees of longitude must be reduced.

For the degrees of latitude one then has

$$x = 4\sqrt{\tfrac{2}{7}} \cdot \sin\tfrac{1}{2}\varepsilon = 2{,}1\ 380\ 900 \sin\tfrac{1}{2}\varepsilon .$$

The appearance of the drawing can be seen in the 20th figure. I will now only also remark that the 17th, 18th, and 20th figures have been drawn to a common scale, and consequently the three parts of the world shown thereon can be compared with each other with respect to their sizes.

Fig. 20.

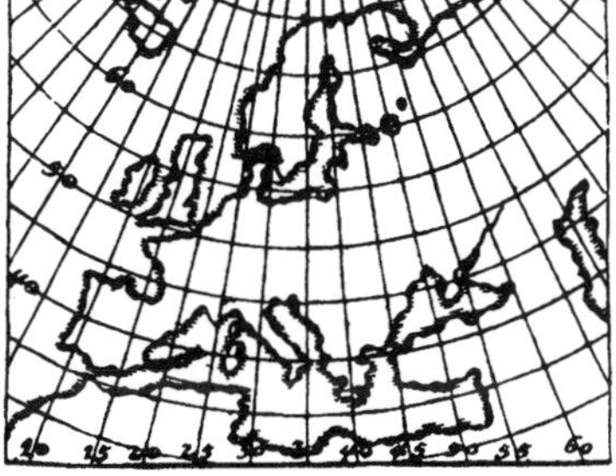

X. REPRESENTATION OF THE SPHEROIDAL EARTH'S SURFACE

§ 111

The earth is generally not very different from a spherical object and consequently one does not pay much attention to its flattened shape in representations of the surface of the earth. A more fundamental consideration is also to be added. It is indeed not possible to represent the earth's surface so that the various parts do not appear dissimilar, since a curved surface cannot be spread out on a plane. The dissimilarities occurring thereby are so appreciable that the differences between the spherical and spheroidal shape of the earth are completely unnoticed. But if these differences are to be considered then the representation must be construed and dedicated to very special points of view.

§ 112

Such a point of view is represented by the use of nautical charts for sailing purposes. Representations that the exact determination of the shape of the earth is important for sailing purposes have not been neglected, and it is indeed worth the trouble to see whether the earth has a large or a small flattening. The improvement of nautical charts and the more exact determination of the direction of ships and their paths depends on this, if ships are to be guided securely in obscuring weather.

§ 113

Special maps may also need to be very exact and a requirement may exist that the ratio of latitude to longitude be exactly as it is on the spheroidal surface of the earth. For the representation of large regions, major parts of the world, or even for the entire surface of the earth, it may be desirable to specify a true ratio along some specified degrees. Finally if it could be shown that it did not require more effort than the representation of a spherical surface then it is apparent that one might as well employ the true figure of the earth. This clearly suggests where our attention should be directed.

§ 114

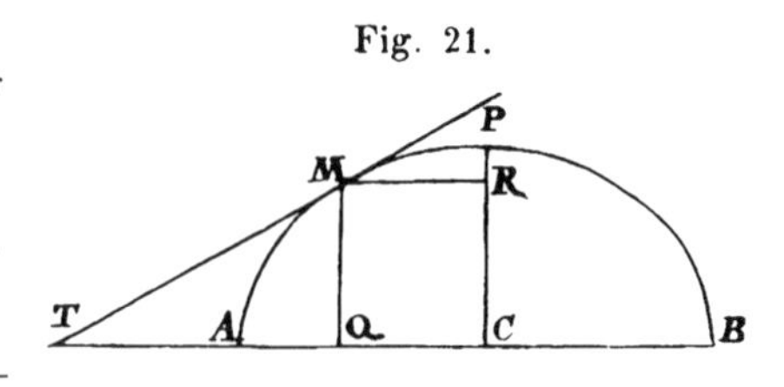

We must first consider the ratio of the degrees of longitude to the degrees of latitude because then most of the material in previous paragraphs will be adaptable without difficulty for the representation of the spheroidal surface. Consequently let $AC = CB$, the radius of the equator, $= 1$, $PC = b$ the semiaxis of the earth, and APB represents a meridian of the earth and which we take to be an ellipse. The eccentricity is

$$e = \sqrt{1 - bb}\ .$$

Further, let M be an arbitrary point on the meridian, MQ is perpendicular to the equator and MR to the axis, MT is a tangent; then the angle TMQ which we set $= p$, is equal to the polar height of M. Additionally, set

$$CQ = MR = x, MQ = CR = y\ ,$$

and the arc

$$AM = v\ ,$$

then from the nature of an ellipse

$$x = \frac{1}{\sqrt{1 + b^2 \cdot \operatorname{tang}^2 p}}\ .$$

Now $x = MR$ is the radius of the parallel circle passing through M. Taking two meridians whose difference in longitude is the differential $d\lambda$, then

$$x\,d\lambda = d\lambda : \sqrt{1 + b^2 \cdot \operatorname{tang}^2 p}$$

is the portion of the parallel through M spanned by these meridians. Refer to this as equal dw, then

$$dw = x\,d\lambda = \frac{d\lambda}{\sqrt{1 + b^2 \cdot \operatorname{tang}^2 p}}\ .$$

§ 115

One finds further, from the character of ellipses, the element of arc of ϖ

$$dv = \operatorname{cosec} p \, dx \,.$$

Now since

$$dx = \frac{b^2 \cdot \operatorname{tang} p \cdot d \operatorname{tang} p}{(1 + b^2 \cdot \operatorname{tang}^2 p)^{3:2}}$$

has been found, then

$$dv = \frac{b^2 \cdot \operatorname{cosec} p \cdot \operatorname{tang} p \cdot d \operatorname{tang} p}{(1 + b^2 \cdot \operatorname{tang}^2 p)^{3:2}} \,.$$

§ 116

From this it follows that

$$dv : dw = \frac{b^2 \operatorname{cosec} p \cdot \operatorname{tang} p \cdot d \operatorname{tang} p}{(1 + b^2 \cdot \operatorname{tang}^2 p)^{3:2}} : \frac{d\lambda}{\sqrt{1 + b^2 \cdot \operatorname{tang}^2 p}}$$

or, reduced to its simplest form

$$dv : dw = b^2 \, dp : \cos p \, (1 - e^2 \sin^2 p) \, d\lambda \,.$$

This is the ratio which was to be found. We now apply it to some representations.

§ 117

On nautical charts the meridians as well as the parallels are represented by straight perpendicularly intersecting lines, and the degrees of the equator are everywhere of the same magnitude with respect to each other as well as with the degrees of every parallel. Therefore

$$dw = d\lambda$$

and consequently

$$dv = \frac{b^2 \, dp}{\cos p \, (1 - e^2 \sin^2 p)} \,.$$

Here v no longer represents the elliptical arc AM but rather its projection on the chart. If this formula is integrated, then

$$v = \tfrac{1}{2}\log\left(\frac{1+\sin p}{1-\sin p}\right) - \tfrac{1}{2}\, e\log\left(\frac{1+e\sin p}{1-e\sin p}\right)$$

or

$$v = \log \operatorname{tang}\,(45^\circ + \tfrac{1}{2}\,p) - \tfrac{1}{2}\, e\log\left(\frac{1+e\sin p}{1-e\sin p}\right)$$

or

$$v = \log \operatorname{tang}\,(45^\circ + \tfrac{1}{2}\,p) - e^2\sin p - \tfrac{1}{3}e^4\sin^3 p - \tfrac{1}{5}e^6\sin^5 p - \text{etc.}$$

In this formula only the first term remains when $e = 0$, the assumption of a spherical earth. One thus sees how the degrees of latitude are diminished in the representation of the spheroidal earth. For brevity set

$$e\sin p = \sin\varphi$$

then one has

$$v = \log \operatorname{tang}\,(45^\circ + \tfrac{1}{2}\,p) - e\log \operatorname{tang}\,(45^\circ + \tfrac{1}{2}\,\varphi)\,.$$

This result is that the direction of a ship with respect to the equator or to a latitude circle is a somewhat smaller angle than on the spherical earth. For example if on the latter a ship travels at an angle of 45° north or south with respect to the equator, then this angle will be about 15 minutes smaller, if in both cases the difference in longitude and latitude is taken to be one degree. On the spheroidal earth the degrees of latitude are furthermore of unequal magnitude and this inequity must also be considered if one wishes to specify everything exactly. The following table is computed according to the first of the formulae.

p	$\log \tan(45^\circ + \frac{1}{2}p)$	$e \log \tan(45^\circ + \frac{1}{2}\varphi)$	v
5	0,0 873 773	0,0 007 560	0,0 866 213
10	0,1 754 259	0,0 015 068	0,1 739 191
15	0,2 648 420	0,0 022 461	0,2 625 959
20	0,3 563 785	0,0 029 686	0,3 534 099
25	0,4 508 754	0,0 036 688	0,4 472 066
30	0,5 493 061	0,0 043 414	0,5 449 647
35	0,6 528 365	0,0 049 814	0,6 478 551
40	0,7 629 098	0,0 055 839	0,7 573 259
45	0,8 813 736	0,0 061 442	0,8 752 294
50	1,0 106 831	0,0 066 579	1,0 040 252
55	1,1 542 346	0,0 071 217	1,1 471 134
60	1,3 169 578	0,0 075 305	1,3 094 273
65	1,5 064 543	0,0 078 825	1,4 985 718
70	1,7 354 151	0,0 081 742	1,7 272 409
75	2,0 275 894	0,0 084 036	2,0 191 848
80	2,4 362 460	0,0 085 688	2,4 276 772
85	3,1 312 900	0,0 086 685	3,1 226 315
90	infinit.	0,0 087 019	infinit.

In this table I have taken, as in the previous treatise,

$$b = \frac{229}{230}$$

and therefore have taken

$$\begin{aligned} e &= 0{,}0\ 931\ 490\,, \\ e^2 &= 0{,}0\ 086\ 767\,, \\ e^4 &= 0{,}0\ 000\ 753\,, \\ e^6 &= 0{,}0\ 000\ 006 \end{aligned}$$

The radius of the equator here is = 1, and thus every degree of the equator is = 0,0 174 533.

§ 118

If this table is expanded to single degrees or even minutes, and if the table of lengths of the degrees given in the earlier treatise is included, then problems occurring in the guiding of a ship are easily solved. I will give only one example. A ship arrives at the 40th parallel from the 30th parallel, having also traversed

10 degrees of longitude. The length of the trip is to be determined on the assumption that the direction was constant. Now in the tables of the previous treatise

from the pole to the 40th degree	2	865	489	units
from the pole to the 30th degree	3	435	205	units
difference		569	716	units

Furthermore, from the most recent table (§ 117)

from the equator to the 40th degree	0,7	573	259
from the equator to the 30th degree	0,5	449	641
difference	0,2	123	612

Finally on the equator 10 degrees = 0,1 745 329. Now set

$$0{,}2\ 123\ 612 : 0{,}1\ 745\ 329 = 1 : 0{,}8\ 218\ 681$$

then 0,8 218 681 is the tangent of the angle of the path of the ship with respect to the meridian. This gives 1,2 943 984 for the secant of this angle, and this value must be multiplied by the units used. The product of 739 013 units is the desired length of the course.

§ 119

We will now return to § 48 and will apply the representation given there to the spheroidal shape of the earth. As there, then, let

$$PM = x\,,$$
$$MN = dx\,,$$
$$MP\mu = m\,d\lambda\,,$$

then (by § 116)

$$MN : M\mu = b^2 dp : \cos p\,(1 - e^2 \sin^2 p)\, d\lambda$$

must hold. But

$$MN = dx, M\mu = x\,m\,d\lambda\,,$$

therefore

$$dx : x\,m\,d\lambda = -\, b^2 dp : \cos p\,(1 - e^2 \sin^2 p)\, d\lambda$$

from which it follows that

$$-\frac{dx}{mx} = \frac{b^2 \cdot dp}{\cos p\,(1 - e^2 \sin^2 p)}$$

and, when one integrates as the earlier section (§ 117) ,

$$\frac{1}{m}\log\frac{1}{x} = \log \operatorname{tang}\,(45^{\circ} + \tfrac{1}{2}\,p) - e \log \operatorname{tang}\,(45^{\circ} + \tfrac{1}{2}\,\varphi)$$

or more briefly

$$\frac{1}{m}\log\frac{1}{x} = v\,.$$

Consequently x is easily determined by means of the last column of the table given in § 117, since $-\log x = mv$. In this formula it is necessary to use hyperbolic logarithms.

§ 120

The value of m can be decided in several ways, as indicated above (§ 48 et seq.). For example, if one takes $m = 1$ then one obtains the method of projection in which the angles at the pole have their true magnitude.

§ 121

But one can also, as in § 51, determine the value of m so that not only $M\mu$ but also $N\nu$ bears the same proportion to MN as is actually found at a given polar height P on the spherical earth's surface. Accordingly

$$\frac{M\mu}{MN} = \frac{mx\,d\lambda}{dx} = -\frac{\cos p\,(1 - e^2 \sin^2 p)\,d\lambda}{b^2 dp)},$$

$$\frac{N\nu}{MN} = \frac{m\,(x + dx)\,d\lambda}{dx}$$

$$^{*}= -\frac{d\lambda\,(\cos p - \sin p\,dp \cdot (1 - e^2 \sin^2 p - 2\,e^2 \sin p \cos p\,dp)}{b^2 dp}\,.$$

---

* [Transl.] This should be corrected to read $= -\,d\lambda \dfrac{\cos p\,(1 - e^2 \sin^2 p) - (1 - e^2) \sin p\,dp}{(b^2 dp)}$, and thus $m = \sin p$.

If the first of these equations is subtracted from the second, then after simplification,

$$m = \sin p - e^2 \sin p\,(1 - 3\cos^2 p)\,.$$

Consequently, using the given polar height P for p

$$m = \sin P - e^2 \sin P\,(1 - 3\cos^2 P)\,.$$

The uses to which this can be put are as in the earlier § 51, 52.

§ 122

The method of representation given in the fifth section (§ 58 and following) can in complete generality also be applied to the elliptical shape of the earth. It is only required that instead of

$$NR : Rr = \lambda \sin \varepsilon : d\varepsilon$$

as in § 61 one uses, as in § 116,

$$NR : Rr = \lambda \cos p\,(1 - e^2 \sin^2 p) : b^2\, dp\,,$$

and, following § 61, set

$$\tfrac{1}{2}\, m\lambda\,(1 - xx) : dx = \lambda \cos p\,(1 - e^2 \sin^2 p) : b^2\, dp\,,$$

whence

$$\frac{2\,dx}{m\,(1 - xx)} = \frac{b^2\, dp}{\cos p\,(1 - e^2 \sin^2 p)}$$

and integrating as in § 117,

$$\frac{1}{m} \log \frac{1 + x}{1 - x} = v$$

is obtained, so that the numbers in the third column of § 117 can also be used here.

§ 123

By means of this formula the surface of the spheroidal earth can be represented in several ways, so that the meridians and parallels are circular arcs, which everywhere cross at right angles, and so that the ratio of degrees of latitude to those of longitude is as actually occurs on the spheroidal surface, and finally so that all angles maintain their true magnitude. If one takes $m = \frac{1}{2}$ then one obtains a representation of the spheroidal earth's surface which is completely similar to that shown in the eleventh figure.

A BIOGRAPHY AND BIBLIOGRAPHY OF JOHANN HEINRICH LAMBERT†

By Hans Maurer, 1931

Genius is not confined to any particular social class, and it makes its own way in the world even under the most unfavorable circumstances. A marvelous example is given by the life of the great mathematician, astronomer, and philosopher, Johann Heinrich Lambert. Descendant of a Palantine family, he was born on August 26th, 1728, the eldest of the children of a poor tailor of Mülhausen in Alsace. At that time this town belonged to Switzerland. The information given below is taken from the book, *Johann Heinrich Lambert, nach seinem Leben and Wirken aus Anlass der zu seinem Andenken begangenen Secularfeier in drei Abhandlungen dargestellt*, published by D. Huber, professor of mathematics, Basle, 1829, on the occasion of the centenary of the birth of Lambert. This centenary was celebrated in 1829 by formal unveiling of a monument at Mülhausen commemorating the life of the scientist. The book contains a biography written by the protestant pastor Graf of Mülhausen, a tribute to Lambert as a philosopher by Professor S. Erhardt of Heidelberg, and a tribute to his memory as a mathematician and physicist by Huber himself.

Lambert's scholastic days ceased at his twelfth year and he was then compelled to work in his father's tailor shop in order to help in supporting his younger brothers and sisters. But the child, eager to learn and whose parents even lacked oil for the evening lamp, used to read at night by the light of the moon or of candles which he obtained by selling drawings, while the others slept. At school he had already learned calligraphy, Latin, and a little geometry; his most prized possession was a book on the science of calculation which he had received from one of his father's customers. From this he learned the calculation of the ecclesiastic calendar and was even able to discover mistakes in the calculations. On the basis of the knowledge acquired from this book he propounded such intelligent questions to the workmen engaged in the wretched shop of his father, that one of them gave the young boy a second volume on arithmetic and geometry. Gradually the attention of other persons was drawn to this infant prodigy.

† From *The Hydrographic Review*, Vol. VIII, No. 1, May 1931, pages 69–82, by kind permission of the International Hydrographic Bureau, Monaco.

Johann Lambert, detail of a lithograph by Gottfried Englemann, after a portrait by Pierre-Roch Vigneron.

A certain Professor Zürcher gave him free instruction in French and the ancient languages and the mayor's clerk Reber procured for him a position as recorder in the mayor's office, later as accountant in a metal factory in Sept, and finally with the professor of law, Iselin, at Basle. The latter permitted Lambert to attend his own lectures on law and also gave him the time to study for himself. Lambert never attended other lectures but, on the other hand, he read all kinds of books: on philosophy by Wolf, Mallebranche, Locke, and on mathematics including algebra and mechanics.

Being later recommended by Iselin, Lambert in 1748 became tutor to the nephew of the Count De Salis and two young men who were related to him. He instructed them in catechism, languages, geometry, military architecture, geography, and history, and meanwhile he himself studied physics, meteorology, mathematics, astronomy, mechanics, metaphysics, and rhetoric from books in the library of the house. He also perfected himself in German, French, Italian, Latin, and Greek languages. As early as 1749 he outlined the plan of his *Lettres Cosmologiques*. In 1750 he commenced his meteorological observations. In 1753 he surveyed the region of Coire and then became a member of the Society for Mathematics and Physics at Basle. His first printed work appeared in the proceedings of this society in 1755.

Beginning in 1756, Lambert undertook several scientific voyages with his pupils. These took him to Göttingen, where he became a corresponding member of the Scientific Society; to Utrecht and The Hague, where in 1758 he published his book on light rays; and to Paris, where he made the acquaintance of D'Alembert; to Marseilles, Nice, Turin, and Milan. In the years that followed we find him in Zürich, Augsburg, Munich, Erlangen, and Leipzig, entering everywhere into close relations with the scientific bodies in those places. In 1764 he arrived in Berlin where, since 1761, he had been a corresponding member of the Academy, and where, eventually, he was to settle down. His first audience with Frederick the Great took the following curious turn:

The King: "Good evening, sir: Will you kindly inform me which of the sciences you have studied in particular?"

Lambert: "All."

The King: "You are also a very able mathematician?"

Lambert: "Yes."

The King: "Which professor instructed you in mathematics?"

Lambert: "Myself."

The King: "You are then a second Pascal?"

Lambert: "Yes, Your Majesty."

Upon which the King returned to his chamber laughing and later, at table, he made the remark that they had proposed for his Academy one of the greatest fools he had ever seen. But about three quarters of a year later, the King expressed his opinion as follows: "In judging this man one should consider only the breadth of his vision and not the minor details." He made him titulary member of the academy with honorary grants, placing him with Euler and three other scientists on the directing committee of the academy, and nominated him chief counsellor for construction, for the general inspection of the public works of the realm. With regard to the latter nomination, Lambert made the following remark to the royal Ministers: "Your Excellencies need not think that I shall revise and correct the ordinary memoranda for the construction works. You may employ clerks for this work if you do not wish to attend to it yourselves. I shall not meddle with questions which concern everybody, as that would be a waste of time for me. If, however, you find difficulties which you cannot solve, then you may call upon me."

We here see Lambert at the apogee of his activity, in active intellectual relations with Euler, Lagrange, Nikolai, Erman, Kant, Mendelssohn, and Bode. Lambert brought about the nomination of Bode as an astronomer at the Berlin Observatory and instigated him to calculate, for many years, the Berlin Ephemerides. If, nowadays, one is inclined to regard Lambert above all as a representative scientist in mathematics and physics, in his day he was also considered a great philosopher whose imposing work, the *Novum Organon* was greatly admired. The high regard in which he was held by Kant is shown by the latter's remark that he would not pass on any proposition which did not conform absolutely to the judgment of Lambert: "If it is not possible to win his assent it is impossible to base this science on incontestable premises."

One can do no better than adhere to the order of precedence which Lambert himself assigned to the mathematicians of his generation: "In the first rank stand Euler and D'Alembert; in the second, Lagrange, who will soon equal the first two; I myself am in the third rank."

The mind of Lambert was constantly engaged in scientific matters without regard for external conditions. In order to study the laws of the reflection of light he simply walked into the finest cafe of Berlin, drew his sword and commenced his studies by making all kinds of movements in front of a large mirror, without taking any notice of the public, who certainly must have considered him a lunatic.

The engraving by Engelmann, reproduced here, after a drawing by Vigneron, gives a good idea of his physiognomy. The face is intellectual and benevolent, and this incited Lavater to write his work on physiognomics. Lambert was a devout Christian and attended church assiduously; his standpoint is contained in his confession of faith: "If Christianity had no mysteries I should doubt it. It would be a poor principle to refuse to believe in anything we cannot conceive; a thing which we are compelled to do daily in so many other matters."

It is characteristic that this same man should think so highly of logical-mathematical reasoning, to which he sought to subordinate everything. If he sought to make general reasoning into a system of mathematics, in which ideas should appear as magnitudes, he demanded that, in morals and in art, the magnitude of every good thing should be comparable with others; or in other words made measurable. He advocated the establishment of a scientific language comprising 106,080 syllables, such that the sense of each word might be logically recognized according to its syllabic construction. At the age of 49 the vital forces of this great man had been exhausted as a result of his incessant labor and following a disease of the respiratory organs an attack of apoplexy put an end to his days on September 25th, 1777.

The prodigious diversity of Lambert's labors in the mathematical and physical sciences is revealed in the amplitude of the bibliographical index given at the conclusion of this article. Many works are of particular interest in the field of activity of the International Hydrographic Bureau. The following remarks might also be of interest.

Among the astronomical works we find alongside of the famous work *Lettres Cosmologiques*, in which he describes the system of the Milky Way of the universe in a manner generally accepted even today, numerous memoranda on comets (at the age of 16 Lambert sought to calculate the path of the comet of 1744), and on the planets; in particular, the presumed satellite of Venus and the reciprocal perturbations of Jupiter and Saturn captivated his imagination. The publication of the Berlin Astronomical Ephemerides was due to his initiative; he contributed to the collection of astronomical tables which the Berlin Academy published in 1779 in three volumes. Science, in recognition of its gratitude, gave the name of Lambert to one of the craters of the moon.

Pure mathematics is indebted to the methods of Lambert which increased the convergence of series, for his arrangement of tables of divisors of numbers, for his method of interpolation as well as his work on the fundamental principles of the calculation of probabilities, and for the Lambert series, which stimulated Lagrange and Laplace in the elaboration of their theories on the development of functions in series. His works on geometry were numerous, both on the laws of perspective and on the quadrature and rectification of curves; they also comprise works which were epoch-making for cartography. He also worked on slide rules and compiled logarithmic and trigonometrical tables.

Physics owes to him the fundamental principles of photometry, the Lambert law of cosines, the conception of the albedo, and his research into the loss of light by reflection. Very justly it has given the name "Lambert" to an absolute unit, that of surface illumination. In the theory of heat, the "Pyrometry" of Lambert already shows the beginnings of the notion of specific heat and his pyramid of colors is the prototype of the "color wheel" of today.

From the point of view of geophysics, we should note the importance of his efforts to serve meteorology: his experiments on the graphical representation of the meteorological elements over a period of time, the variation of the barometric height with the seasons and the movements of the moon, the calculation of the average wind over an interval of time and attempts to develop hygrometry. From the standpoint of terrestrial magnetism, it is interesting to note a memorandum by Lambert on the point of intersection of the isogon of 15° in Africa on the isogonic chart for the year 1770. This point probably corresponds to the one which is now in the Atlantic, from which point the magnetic variation increases towards the north and south and decreases toward the west and east.

The astounding mass of work accomplished by this mathematical genius, which applies directly to the affairs of everyday life, may be readily seen from the list of his works. Among them we find articles on lighting apparatus, ink and paper, windmills and watermills, four-wheeled vehicles, beds for invalids, and bellows.

For geographers and seamen, Lambert's cartographic work was extremely fruitful. We find these principally in the memoirs entitled: *Anmerkungen und Zusätze zur Entwerfung der Land und Himmelskarten* (Notes and addenda on the projection of terrestrial and celestial charts). In this memorandum Lambert compares the different projections with each other and adds other important ones of his own. These may be briefly classified as follows:

For the equidistant azimuthal projection, which is more often attributed to Postel, but which had been employed even before by Gerhard Mercator, Lambert calculated the tables giving the azimuth and the distance from the center of the chart for every five degrees of latitude and longitude.

The equivalent azimuthal projection originated with Lambert himself. The law of the radius in this is $\rho = 2r\sin\frac{\delta}{2}$, in which r is the radius of the sphere and δ the angular distance from the center of the chart.

Among the cylindrical projections the equivalent projection was invented by Lambert. For a position of the axis coinciding with the axis of the earth, the law for the scale at each meridian, for latitude φ, is $y = r\sin\varphi$.

For a transverse position of the axis (axis of the cylinder perpendicular to the axis of the earth) it is called the Lambert transverse isocylindric projection. Lambert calculated a table for the rectangular coordinates x and y, from the formulae:

$$\cot\frac{x}{r} = \cot\varphi\cos\lambda\,;\, y = r\cos\varphi\sin\lambda\,.$$

The conformal cylindrical projection was also described by Lambert, under the name Lambert conformal cylindrical projection, for a transverse position of the axis. Whereas, where the axis of the cylinder lies in the axis of the earth, it represents the Mercator projection, the equations for the coordinates for the transverse position (assuming $r = 1$) become:

$$\cos\varphi\cos\lambda = \frac{e^{y} - e^{-y}}{e^{y} + e^{-y}}\,;\,\cot\varphi\cos\lambda = \cot x\,.$$

The loxodromes (rhumbs), which on Mercator's chart are straight lines, are not so on this projection.

In the ordinary conical projection, in which the axis is oriented in line with the earth's axis, the meridians become straight lines at angles $u = m\lambda$ where λ is the difference in longitude and m is a constant, while the parallels of latitude and usually the pole also, are the arcs of concentric circles. Lambert has also proposed an equivalent projection of this type (also called the isospheric stenoteric conical projection) in which the pole is represented by a point.

Besides the equation $u = m\lambda$ for the lie of the meridians, the law of the radius may be applied to each meridian:

$$\rho = \frac{2r}{\sqrt{m}} \sin\left(45 - \frac{\varphi}{2}\right);$$

r is the radius of the sphere and φ = geographic latitude.

If the angles are to be conformal throughout the length of the parallel of latitude φ_o we should take

$$m = \cos^2\left(45 - \frac{\varphi_o}{2}\right)$$

Lambert also developed the formulae for the ellipsoid: for a semimajor axis A and the eccentricity ε we have:

$$\rho = 2A\sqrt{\frac{1-\varepsilon^2}{m}} \sin\left(45 - \frac{\varphi}{2}\right)\left[1 + \frac{\varepsilon^2}{3}(1 + \sin\varphi + \sin^2\varphi)\right]$$

For the conformal conical projection, also discovered by Lambert, the formulae to be applied are:

$$u = m\lambda\,; \rho = a\left[\operatorname{tang}\frac{p}{2}\cot\frac{p_o}{2}\right]^m; k = \frac{m\rho}{r\sin p}.$$

Here ρ is the distance of the apex of the cone for the polar distance p; p_o, a constant, the original polar distance, r the radius of the sphere, m and a are constants and k is the scale ratio between the chart and the sphere for the polar distance p.

If the cone is to touch the sphere at the polar distance p we must take:

$$m = \cos p_o\,; a = r\tan p_o\,.$$

We can also determine m and a in such a manner that $k = 1$ on two parallels of polar distances p_1 and p_2. We should then have:

$$m = \frac{\log\sin p_1 - \log\sin p_2}{\log\operatorname{tg}\frac{p_1}{2} - \log\operatorname{tg}\frac{p_2}{2}}.$$

The chart may then be considered as having been constructed for a cone which cuts the sphere at the parallels at polar distances p_1 and p_2, but also for a cone tangent to the sphere at the circle having a polar distance of $p_o = \text{arc} \cos m$. The scale constant $\frac{a}{r}$ only is differently conceived in the two cases.

For $p_o = 0$ the cone becomes reduced to the plane tangent to the pole and the projection becomes stereographic in accordance with the laws:

$$u = \lambda\,; \rho = 2\,r\,\text{tg}\,\frac{p}{2}$$

For $p_o = 90^o$, it becomes the Mercator projection. The complicated equations for the conformal conical projection for the ellipsoid will not be cited here.

The Lambert conformal conical projection represents the general case of a conformal projection with the meridians as straight lines; Lambert pointed out also the still more general case of the conformal projection in which the meridians and the circles of latitude are orthogonal circles (including the limiting case, in which they become straight lines). Ordinarily this projection is called the projection of Lagrange who published a description of it in the *Memoirs of the Academy of Berlin* in 1781; but it had already been reported by Lambert in 1772.

In this projection the meridians form a group of circles passing through the representations of the two poles and intersecting each other at angles $m\lambda$, where λ is the difference of longitude. The representation of the circles of latitude is provided by the corresponding group of orthogonal circles. In a system of rectangular coordinates (x, y) of which the origin is one of the poles and the axis of the ordinates is the central meridian, the following formulae apply:

$$x = \frac{-A \sin m\lambda}{N}\,;\ y = \frac{A \cos m\lambda + B \cot^m \left(45 + \frac{\varphi}{2}\right)}{N}\,;$$

$$N = A^2\,\text{tg}^2 \left(45 + \frac{\varphi}{2}\right) + 2\,AB \cos m\lambda + B^2 \cot^m \left(45 + \frac{\varphi}{2}\right)$$

in which A, B and m are constants; $\frac{1}{B} = 2\,b$ is the distance of the representations of the poles on the rectilinear meridian, $\lambda = 0$. If we take $\frac{A}{B} = \alpha$, the coordinates of the centre of the circle representing the meridian λ are $y = b$; $x = b \cot m\lambda$ and its radius $\rho = b\,\text{cosec}\, m\lambda$.

The coordinates of the centre of the circle representing the parallel of latitude φ are

$$x = 0, y = \frac{2b}{\alpha^2 \operatorname{tg}^{2m}\left(45 + \frac{\varphi}{2}\right) - 1},$$

and its radius

$$\rho = \frac{2\alpha b \operatorname{tg}^{m}\left(45 + \frac{\varphi}{2}\right)}{\alpha^2 \operatorname{tg}^{2m}\left(45 + \frac{\varphi}{2}\right) - 1}.$$

If the parallel φ is to be a straight line, then we must have

$$\alpha = \cot^{m}\left(45 + \frac{\varphi}{2}\right).$$

b, φ_o and m are therefore still available.

$\varphi_o = 0$ gives the charts with rectilinear equator. If $m = 1$ the result is a transverse stereographic projection; while with $m = \frac{1}{2}$ a chart is produced which gives the whole world conformal within a circle.

If we assume that one of the poles is moved to infinity; i.e., if we assume $B = 0$ and $A = 1$, we obtain the conical conformal projection of Lambert with rectilinear meridians and, in its limiting cases, the Mercator projection in one case and, in the other, the stereographic polar projection. We see therefore in the Lambert projection with groups of orthogonal circles a form which is the general origin of conformal projections. The Lambert groups of orthogonal circles are also introduced into other conformal charts of a quite different nature. In the Littrow (1833) conformal chart, where the meridians and the parallels of latitude form a system of hyperbolae and homofocal ellipses, in which all the straight lines are lines of equal azimuth, and of which the identity with the curves of the Weir azimuth diagram was proved by the author of this article in 1905, the system of arcs of great circles passing from the centre of the chart, through points on the equator 90° apart, as well as that of the horizontal circles described about them, is a system of groups of Lambert orthogonal circles, as shown by A. Wedemeyer in 1918. Consequently the Littrow chart is also a chart with the Lambert system of circles but conceived from another point of view.

More than 200 years have passed since this great genius appeared on earth. Even today we are filled with admiration for what this absolutely self-taught man was able of himself to produce during his short existence. Endowed with the faculty of submitting practically all questions in which he was interested to mathematical treatment and to plumb them by means of systematically ordered thought, we find him in the most varied fields, and always as a pioneer in new scientific conceptions.

We do not wish to close this short sketch of the life of this great man without pausing an instant to consider his magnanimity. Always taking for granted the benevolence of the Creator, he advanced the audacious hypothesis, in his *Lettres Cosmologiques*, that the comets of our system must be inhabited by happy creatures. In accordance with the motto of creation recognized on earth "As much life as possible," the existence in the solar system of as many comets as there are stable trajectories, the number of which Lambert calculated as approximately 12,000 must be assumed. This conception corresponds to the maxim which he wrote in his own hand which we find below the appended portrait. The translation thereof is as follows: "On our earth, organic bodies are, among all others, those which are created most abundantly and most easily.... Everything in this world for which the means are most abundantly provided, must be considered as being part of the scheme of creation..."

The inscription on the monument of Lambert at Mülhausen is:

Johannes Henricus Lambert
natus Mulhusii 26 Aug. 1728, denatus Berolini 25 Sept. 1777.
Dem durch Selbsttätigkeit entwickelten grossen Geiste.
Ingenio et studio
Sa cendre repose á Berlin, son nom est éscrit dans les Fastes d'Uranie.

THE WRITINGS OF JOHANN HEINRICH LAMBERT

I. Works published separately

1. Les propriétés remarquables de la route de la lumière par les aires, et en general par plusiers millieux réfringens, sphériques et concentriques. La Haye 1758. [On the path of light in refractive media].
2. Die freie Perspekive oder Anweisung, jeden perspekivischen Aufriss von freier Stücken und ohne Grundriss zu verfertigen. Zürich 1759 [On free perspective; 2nd edition with additions, Zürich. 1774].
3. Photometria sive de mensure et gadibus luminia colorum et umbriae. Augustae Vindelicorum, 1760. [On photometry].
4. Insigniores orbitae cometarum propritates. Aug. Vind.. 1761. [On the orbits of comets].
5. Cosmologische Briefe über die Einrichtung des Weltbaues. Augbsburg. 1761. [Cosmological letters. Also published as "Systèm du Monde, par Lambert," Mérian, Berlin, 1770].
6. Beschreibung und Gebrauch der logaritmetischen Rechenstäbe. Augsburg, 1761. [Description and use of logarithmic slide rules].
7. Neues Organon oder Gedanken über die Erforschung u. Bezeichnung des Wahren und dessen Unterscheidung vom Irrtum und Schein; 2 parts, Leipzig, 1764. [How to distinguish truth from error and appearance].
8. Beiträge zum Gebrauche der Mathematik u. deren Anwendung; 3 parts in 4 volumes. Berlin 1765, 1770, 1772. [Use and application of Mathematics. Part III. contains the work on map projections]:

   Part I:
   1. Anmerkungen u. Zusätze zur prakischen Geometrie. [Practical geometry].
   2. Die Visierkunst, sowohl ganz, als nicht ganz angefüllt liegende Fässer. [The art of sighting with full and partially filled vessels].
   3. Anmerkungen u. Zusätze zur Trigonometrie. [On Trigonometry].
   4. Theorie der Zuverlässigkeit der Beobachtunen und Versuche. [The reliability of observations and experiments].

   Part IIa:
   1. Teilung u. Teiler der Zahlen. [Divisors and parts of numbers].
   2. Vorschlag die Teiler der Zahlen in Tabellen zu bringen. [Proposal for Tables of the parts of numbers].
   3. Verwandlung der Brüche. [Conversion of Fractions].
   4. Algebraische Formeln für die Sinus von 3 zu 3 Graden. [Formulae for sines by three degrees].
   5. Vorläufige Kenntnisse für die, die Quadratur und Rektifikation des Circlus suchen. [Preliminary advice for the squarers of circles].
   6. Einige Anmerkungen von Ausmessung der Winkel und Linien auf dem Papier. [Measurement of angles and lines on paper].
   7. Anlage der Tetragonomtrie. [On tetragonometry].
   8. Anmerkungen über die Verwandlung und Auflösung der Gleichungen. [Transformation and solution of equations].

9. Quadratur u. Rektification der krummen Lienien durch geradlinige Vielecke, welche dieselben und in denselben beschrieben warden können. [Squaring and rectification of curved lines using polygons].
10. Anmerkungen und Zusätze zur Gnomonik. [On gnomonitry].

Part IIb:

11. Gedanken über die Grundlehren des Gleichgewichts und der Bewegung. [Thoughts on the fundamentals of gravity and motion].
12. Zergliederung u. Anwendung der Mayerschen Mondstaflen, u.s.w. [Use of Lunar tables].

Part III:

1. Eine besondere Eigneschaft der tangenten. [A special property of tangents].
2. Zusätze zur Visierkunst. [On sighting].
3. Rektifikation elliptischer Bogen durch unentiche Reihen. [Rectification of elliptical arcs using infinite series].
4. Verwandlung der Figuren in gleich grosse Rektangel. [Conversion of figures to rectangles of the same size].
5. Anmerkungen über das Einschalten. [On interpolation].
6. Anmerkungen and Zusätze zur Entwerfung der Land- und Himmels-Karten. [On maps].
7. Von Beobachtung und Berechnung der Cometen, u. besonders des Cometen von 1769. [On comets].
8. Anmerkungen über die Baukunst. [On construction].
9. Anmerkungen über die Sterblichkeit, Totenlisten, Geburten und Ehen. [Mortality, life tables, births and marriages].
10. Beschreibung u. Gebrauch einer neuen and allgemeinen elliptischen Tafel, worauf alle Finsternisse des Mondes und der Erde vorgestellt werden, u.s.w. Berlin 1765. [Use of a new elliptical table].
11. Anmerkungen über die Gewalt des Schiesspulvers u. den Wierderstand der Luft u.s.w. Dresden, 1766. [Strength of gun powder and air resistance].
12. Anmerkungen über die Brander'schen Mikrometer von Glas u.s.w. Augsburg, 1769. [Micrometers of glass].
13. Kurzgefasste Regeln zu perspektivischen Zeichnungen vermittelst eines zu deren Ausübung eingerichteten Proportional-Cirkels. Augsburg, 1768. [Perspective drawing using a proportional compass].
14. Picards Abhandlung vom Wasservägen. Berlin 1770. [Water levels].
15. Zusätze zu den logarithmischen u. trigonometrischen Tabellen, etc. Berlin 1770. [Addenda to logarithmic and trigonometric tables].
16. Anlage zur Architektonik oder Theorie des Einfachen and Ersten in der philosophischen and mathematischen Erkenntniss. 2 Vol. Riga 1771. [Theory of simplicity; also as “Exposé des points fondamentaux de la doctrine des principles de Lambert”; la Haye 1780, by Tremblay].
17. Beschreibung einer mit Calauischen Wachse ausgemalten Farben-Pyramide, wo die Misdung jeder Farbe aus Weiss and 3 Grundfarben angeordnet wird; Berlin 1772. [Mixing of colors using primaries].

9. Pyrometrie oder vom Masse des Feuers and der Wärme; Berlin 1779. [Heat content of fire and warmth].
10. Poetische Beschreibung der Aussicht der Gegend um Chur. In J. Bernouilli's Sammlung Kurzer Reisebeschreibungen; part II, 1781. [Poetic description of the vicinity of Chur].
11. Deutscher gelehrter Briefwechsel, J. Bernouilli, 5 Vol. Berlin 1782-84. [Correspondence].
12. Logische und philosophische Abhandlungen; J. Bernouilli, 2 parts, Dresden 1782. [Logical and philosophical works].
13. Fünf philosophische Briefe Lamberts u. Kants, in Immanuel Kant's vermischten Schriften Vol. II; Halle 1799. [Five philosophical letters of Lambert and Kant].

II. Treatises Published In Memoirs

A. *Mémoires de 1'Académie Royale des Sciences de Berlin.*

1761 (1768):

1. Mémoires sur quelques propriétés remarquables des quantities transcendantes, circulaires et logarithmiques. [Remarkable properties of transcendant, circular, and logarithmic quantities].
2. Expériences sur le poids du sel et la gravité specifique des saumures. [Weight of salt and specific gravity of brine].
3. Sur la méthode du calcul intégral. [On integral calculation].

1763 (1770):

4. Sur quelques instruments acoustiques. [Acoustical instruments, German edition with additions by Prof. Huth; Berlin 1796].
5. Observations sur les équations d'un degré quelconque. [Equations of arbitrary degree].
6. Observations sur les diviseurs d'un degré quelconque, qui peuvent être trouvés indépendamment de la solution des équations. [Divisors of arbitrary degree that can be found without solving the equations].
7. Observations sur quelques dimensions du monde intellectual. [Dimensions of the intellectual world].

1765 (1767):

8. Mémoire sur la résistance des fluides avec la solution du probleme ballistique. [Resistance of fluids solving ballistics problems].
9. Discours de reception de Mr. Lambert comme Membre de 1'Académie (10.1 1765). [Reception on entrance to the Academy].

1766 (1768):

10. Analyse de quelques expériences faites sur 1'aimant. [Analysis of experience with magnets].

1767 (1769):

11. Sur la courbure du courant magnétique. [Curvature of magnetic currents].
12. Sur la figure de l'Océan. [On the shape of the ocean].
13. Solution générale et absolue du probléme de trois corps, moyennant des suites infinies. [Solution of the three body problems, using series].

1768 (1770):

14. Sur la vitesse du son. [On the speed of sound].
15. Mémoire sur la partie photométrique de tout l'art de peindre. [Photometry and painting].
16. Observations trigonométriques. [Trigonometric observations].

1769 (1771):

17. Essai d'hygrométrie, ou sur la mesure de 1'humidité. [On measuring humidity].

1770 (1772):

18. Quelques remarques sur la cométe de 1769. [The comet of 1769].
19. Sur les porte-lumiéres appliquées à la lampe. [On lighting].
20. Observations sur 1'encre et le papier. [On ink and paper].
21. Observations analytiques. [Analytical observations].
22. Essai de taxeometrie, ou sur la mesure de 1'ordre. [Essay on taxonomy, or the measurement of order].

1771 (1773):

23. Exposé de quelques observations qu'on pourroit faire pour répandre du jour sur la météorologie. [Meteorological observations to pass the time].
24. Observations sur 1'influence de la lune dans le poids de 1'atmosphére. [The moon's influence on the weight of the atmosphere].
25. Sur les lorgnettes achromatiques d'une seule espéce de verre. [Achromatic lenses from one piece of glass].
26. Observations sur 1'orbite apparente des cométes. [On the apparent orbit of comets].
27. Examen dune espéce de superstition ramenée au calcul des probabilités. [Probablistic calculations about a superstition].

1772 (1774):

28. Sur le frottement, en tant qu'il ralentit le mouvement. [On polish, and when it retards movement].
29. Sur la fluidité du sable, de la terre et d'autres corps mous, relativement aux lois de 1'hydrodynamique. [On the fluidity of sand, and other earth materials, relative to the laws of hydrodynamics].
30. Suite de 1'essai d'hygrométrie. [More on humidity; German translation by Tenn in Nr. 17; Augsburg 1774].
31. Sur la densité de l'air. [On the density of air].

1773 (1775):

32. Rapport fait à 1'Académie au sujet des six traités de M. De Nase. [Report to the Academy on Nase's six works].
33. Construction d'une échelle ballistique. [Construction of a ballistic scale].
34. Exposé de quelques observations physiques. [Report of physical observations].
35. Résultat des recherches sur les irrégularites du mouvement de Saturne et de Jupiter. [Result of investigations on the irregularities of movement of Saturn and Jupiter].
36. Essai d'une théorie du satellite de Vénus. [On the theory of the satellite of Venus].
37. Second essai de taxéométrie, ou sur la mesure de 1'ordre. [Second essay on taxonomy].

1774 (1776):

38. Rapport fait à 1'Académie au sujet du'un manuscrit du R. P. Knoll (sur un lit pour malades). [Report to the Academy on Knoll's manuscript on a bed for the ill].
39. Remarques sur le tempérament en musique. [On temperament in music].
40. Sur la perspective aérienne. [Aerial perspective].
41. Observations sur les flûtes. [On flutes].
42. Expériences et remarques sur les moulins, que 1'eau meut par en bas. [Experiences with mills which raise water].
43. Remarques sur les moulins et autres machines dans les roues qui prennent 1'eau à une certaine hauteur. [Mills which raise water to a certain level].
44. Remarques sur les moulins et autres machines où 1'eau tombe en dessus de la roue. [Mills which use falling water].
45. Remarques sur les moulins à vent. [Windmills].

1776 (1779):

46. Sur le frottement, en tant qu'il ralentit le mouvement et s'y oppose. Second mémoire. [On polish, and when it aids and retards movement, 2nd study].
47. Sur les forces du corps humain. [On the forces of the human body].

1777 (1779):

48. Sur les observations du vent. [Observations on wind].
49. Avertissement de M. Bernouilli concernant les deux mémoires suivants. [Bernouilli's remarks on the following two papers].
50. Sur les irrégularités du mouvement de Saturne. [Irregularities in the movement of Saturn].
51. Sur les irregularites du mouvement de Jupiter. [of Jupiter].

1783 (1785):

52. Sur le quarré de la vitesse dans la dynamique. Communiqué par M. Bernouilli. [On the square of the speed in dynamics].

1784 (1786):

53. Avertissement de M. Bernouilli sur le mémoire suivant. [Bernouilli's remarks on the following].
54. Sur les fluides considérés relativement à 1'hydrodynamique. [Fluids considered in relation to hydrodynamics].

B. *Abhandlungen der Churfürstlich Bayerischen Akademie der Wissenschaften.*

Vol. I. Munchen 1763:

1. Abhandlung von dem Gebrauch der Mittagslinie beim Land-und Feldmessen. [Use of the meridian in surveying].
2. Abhandlung von den Barometer-Höhen and ihren Verānderungen. [Barometric heights and their variations].

C. *Actis Helveticis physico-mathematico-anatomico-botanico-medicis.*

Vol. II. Basileae 1755:
1. Tentamen de vi caloris, qua corpora dilatat ejusque dimensione.

Vol. III. Basileae 1758:
2. Theoria staterarum ex principiis mechanices universalif exposita.
3. Observationes variae in mathesin puram.
4. Observationes meteorologicae curiae Rhaetorum habitae.

Vol. IV. Basileae 1760:
5. De variationibus altitudinum barometricarum a luna pendentibus.

Vol. IX. Basileae 1787:
6. Sur le son des corps elastiques. [The sound of elastic bodies].
7. Sur les machines, qui produisent leur effet au moyen d'une manivelle. [Machines which produce their effect by a crank].

D. *Berliner astronomisches Jahrbuch oder Ephemeriden.*

1774 (1776):
1. Über das Einschalten beim Gebrauche der Ephemeriden. [Interpolation in ephemerides].
2. Über die Nutation. [On nutation].
3. Über die Abirrung des Lichtes der Planeton and der Fixsterne. [Deviation of the light from planets and fixed stars].
4. Von der Parallaxe and dem Durchmesser des Mondes in verschiedenen Höhen. [Parallax and the diameter of the moon at various altitudes].
5. Von der scheinbaren Gestalt des Ringes des Saturns. [On the apparent form of Saturn's rings].
6. Erklärung and Gebrauch der Mond-Karte. [Explanation and use of lunar maps].
7. Vom Auf-und Untergang des Mondes and dessen Bestimmung für jede Oerter der Erdfläche vermittelst der Ephemeriden. [Rising and setting of the moon, by place].
8. Vom Gebrauche der Ephemeriden bei Mond-Uhren. [Using ephemerides and moon clocks].
9. Anmerkung über P. Hallensteins Bestimmung des Meridian Unterschieds zwischen Peking und Petersburg. [Remarks on the longitudinal difference between Peking and Petersburg by P. Hallenstein].
10. Vom Gang der Pendeluhren. [Pendulum clocks].

1775 (1777):
11. Vom Gebrauche der Mond-Karte bei Sternbedeckungen. [Using lunar charts during star obscuration].
12. Von der geographischen Länge and Breite der Oerter. [The geographic longitude and latitude of places].
13. Von Bestimmung und Berichtigung der Mittagslinie. [Determination and correction of the meridian].

14. Über die neuen Versuche, das Feld der Fernrohre zu erweitern. [On the attempts to enlarge the field of view of telescopes].
15. Betrachtung über die Monds-Finsternis vom 30. Sept. 1778. [Report on the lunar eclipse.]
16. Von den Cometen 1773 und 1774. [The comets of 1773 and 1774].
17. Betrachtungen über die veraenderliche Sichtbarkeit des Saturn-Ringes. [Observations on the variable visibility of Saturn's rings].
18. Über die scheinbare Lage der Trabanten des Saturns. [On the apparent orbit of Saturn].
19. Nachricht von des gegensseitigen Störungen des Jupiters and Saturns. [The mutual influence of Jupiter and Saturn].
20. Vom Trabanten der Venus. [The path of Venus].
21. Von den Grenzen der Moeglichkeit der Sonnen-Finsternis and Sternbedeckungen vom Monde. [Bounds on the possibility of obscuration of the sun and star coverages of the moon].
22. Von einer neuen Art Sonnenuhren. [A new type of sun dial].

1776 (1778):

23. Neue Art Sonnenfinsternisse zu entwerfen. [A new method of obtaining sun obscurations].
24. Scheinabre Lage der Saturns-Trabanten im Jahre 1778. [Apparent position of Saturn's path in 1778].
25. Über die Anwendung der Aequilibrations-Linie bei Mauer-Quadranten. [Use of equilibrium lines on Maurer's quadrats].
26. Von der Sichtbarkeit des Saturn-Ringes (Fortsetzung). [Visibility of Saturn's rings, con't].
27. Vom Trabanten der Venus. [The path of Venus].
28. Einige trigonometrische Anmerkungen. [Some trigonometric remarks].
29. Einige Anmerkungen ueber die Kirchen-Rechnung. [Some remarks on church calculations].

1777 (1779):

30. Über die Bedeckung des Jupiters vom verfinsterten Monde. [Coverage of Jupiter by the obscured moon].
31. Erklärung der magnetischen Abweichungs-Karte. [Explanation of the map of magnetic deviations].
32. Einige Anmerkungen über die Uhren. [Some remarks on clocks].
33. Gebrauch der Monds-Karte bei der Mondsfinsternis den 17. III. 1764. [Use of the lunar chart during the obscuration of the moon on March 17, 1764].
34. Über die Bestimmung der Laufbahn der Cometen. [Determination of the path of comets].
35. Anmerkungen über Strahlenbrechung. [Remarks on calculation of rays].

1777 (1780):

36. Anmerkungen über die Zeitgleichung. [Remarks on the equation of time].
37. Fortgesetzte Anmerkungen ueber den Gang der Wollastonischen Uhr. [Continued remarks on the Wollastish clock].
38. Bedingungen ganzer Sonnenfinsternisse für eine gegebene Pol-Höhe. [Conditions for a total eclipse at a given zenith altitude].

39. Anmerkungen and Aufgaben zum Gebrauche des in den Ephemeriden engegebenen Mondlaufes. [Use of the lunar path as given in ephemerides].
40. Bemerkungen über die nahen zusammenkünfte der Planeten. [Remarks on the near approaches of the planets].
41. Über die groesste Abweichung der untern Planeten. [On the greatest separation of the lower planets].
42. Vom Glanze der Venus. [On Venus's shine].
43. Über die Umgälzung der Sonne um ihre Achse. [On the rotation of the sun on its axis].
44. Analytische Formeln zum Behufe der astronomischen Rechnung. [Analytical formulae helpful in astronomy].
45. Zusatz zu Lehre vom Einschalten. [Contribution to interpolation studies].
46. Über einen besonderer Gebrauch der Epheriden. [A special use of ephemerides].

1778 (1781):

47. Anmerkungen über den Positions-Winkel des Mondes. [The position angle of the moon].
48. Zur Bestimmung der Zeit, wenn zwei Sterne in gleichen vertikalichen Kreis kommen. [Determination of the time at which two stars reach the same vertical circle].
49. Sammlung astronomischer Tafeln, unter Aufsicht der Kgl. Preussischen Akademie der Wissenschaften; 3 volumes Berlin 1776. [Collection of astronomic tables by Lambert].

E. *Novis actis eruditorum Lipsiensibus.*

1. De ichnographica campi vel regionis delineatione independenter ab omni basi perficienda. 1763.
2. De universaliori calculi idea, cum annexo specimine. 1764–65.
3. In algebram philosophicam et Richeri breves annotationes. 1766–67.
4. De topicis schediasma. 1768.
5. Adnotata quaedam de numeris eorumque anatomia. 1769.
6. Solutio problematis ad methodum tangentium inversam pertinentis. 1769.

F. *Leipziger Magazin fuer reine and angewandte Mathematik*, edited by J. Bernouilli and C. F. Hindenburg.

1786: 2nd part:

1. Theorie der Parallel-Linien. [Theory of parallel lines].

1786: 3rd part:

2. Fortsetzug über die Parallel-Linien. [Continuation on parallel lines].

1786: 4th part:

3. Anmerkung über die Bestimmung des körperlichen Raumes jeder Segmente, welche durch die Umdrehung einer konischen Sektion entstehen. [Determination of the volume of segments obtained by rotation of a conic section].

1786: 1st piece:

4. Über die Mehrheit der Wurzeln hoeherer Gleichungen. [On the multiplicity of roots of higher equations].

1786: 3rd piece:

5. Fernere Anwendung der Mayer'schen Mondtafeln. [Further application of lunar tables].

1786: 4th piece:

6. Differental-und Integral-Rechnung endlicher Grössen. [Differental and integral calculation of finite quantities].

G. *Archiv der reinen and angewandten Mathematik*, edited by Hindenburg.

1796. 5th book:

1. Über die vierraedrigen Wagen. [On four wheeled vehicles].

1798. 7th book:

2. Über die Bewegung der Faesser, in welchen Kugeln geründet werden. [The movement of vessels in which spheres are rounded].

1799. 9th book:

3. Grundsätze der Perspektive, aus Betrachtung einer perpektivisch gezeichneten Landschaft abgeleitet. [Fundamentals of perspective, derived from a perspectively drawn landscape].
4. Optische Betrachtungen (Über den Ort des Bildes bei Spiegeln). [Optical considerations—the image of a picture in a mirror].

1799. 10th book:

5. Versuche and Berechnungen ueber die Blasebaelge. [Experiments and calculations on bellows].
6. Mathematische Ergötzungen über Glücks spiele. [Mathematical delights in games of chance].

Lambert monument in its present location in Mülhouse, in the Alsace region of eastern France, near the borders with Switzerland and Germany. Photograph by W. Tobler, July 1972.

SELECTED REFERENCES

d'Avezac, M., "Coup d'oeil historique sur la projection des cartes de géographie," *Bulletin*, Societé Géographique, Paris, 5, 1863.

Albers, H., "Beschreibung einer neuen Kegelprojection," *Zach's Monatliche Corresondenz*, pp. 97–114, 240–250, 450–459, 1805.

Berggren, J. L., A. Jones, *Ptolemy's Geography: An Annotated Translation of the Theoretical Chapters*, Princeton University Press, 2000.

Biernacki, F., *Theory of Representation of Surfaces for Surveyors and Cartographers*, U.S. Department of Commerce translation from the Polish, Springfield, 1965.

Blome, R., *Cosmography and Geography; the first part being A Translation from that Eminent and much Esteemed Geographer, Varenius, Roycroft*, London, pp. 318 et seq., 1693.

Boole, G., *The Laws of Thought*, Dover reprint, New York 1958.

Brandenberger, C., *Verschiedene Aspekte und Projektionen fuer Weltkarten*. Doctoral dissertation, Institut für Kartographie, ETH Zurich 1996.

Brown, B. H., "Conformal and Equiareal World Maps," *American Mathematical Monthly*, 42: 1–80, 1935.

Bugayevskiy, L. M.; Snyder, J. P., *Map Projections:* A *Reference Manual.* Taylor & Francis, London, 1995.

Canters, F.; Decleir, H., *The World in Perspective.* J. Wiley, Chichester, 1989.

Canters, F., *Small-Scale Map Projection Design*, London, Taylor and Francis, 2002.

Chebyshev, P. L., "Sur la construction des cartes géographiques," Académie Impériale des Sciences, Physico-Mathématique, *Bulletin*, Vol. 14, 1856: 257–261. Also: Oeuvres de P. L. Tschebyshev, Vol 1., Chelsea, New York.

Daly, C. P., "On the early history of cartography, or what we know of maps and map-making before the time of Mercator," *Journal* of the American Geographical Society, XI: 1–40, 1879.

Dorling, D. Newman, M., Barford, A., *The Atlas of the Real World*, Thames & Hudson, New York, 2008.

Euler, L., "Drei Abhandlungen über Kartenprojection," A. Wangerin, ed., *Ostwald's Klassiker der exakten Wissenschaften*, 93, Leipzig, Engelmann, 1898.

Eves, H., *Great Moments in Mathematics (Before 1650)*, The Mathematical Association of America, Pridence, 1980.

Fenna. D., *Cartographic Science*, London, Taylor & Francis/CRC, 2007.

Fiala, F., *Mathematische Kartographie*, translated from Czech, Berlin,VEB, 1957.

Gauss, C. F., "Allgemeine Auflösung der Aufgabe: Die Theile einer gegebnen Fläche auf einer andern gegebnen Fläche so abzubilden, dass die Abbildung dem Abgebildeten in den kleinsten Theilen ähnlich wird, " A. Wangerin, ed., *Ostwald's Klassiker der exakten Wissenschaften*, 55, Leipzig, Engelmann, pp. 57–101, 1894.

Grave, D.A., "Sur la construction des cartes géographiques," *Journal des mathématiqes pures et appliqués*, series 5, Vol.2: 317–361, 1896.

Günther, S., E. Hammer, and H. Haack, "Die Fortschritte der Kartenprojektionslehre," Geographisches Jahrbuch, IX, X, XII, XIV, XVII, XIX, XX, XXIV, XXVI, XXIX, XXXIII, XLVI, LI, LVII, 1882–1942.

Hessler, J., *Projecting Time: John Parr Snyder and the Develoment of the Space Oblique Mercator Projection*, Philip Lee Phillips Society, Occasional Paper Series, No. 5, Library of Congress, Washington, D.C., 2004.

Hoschek J., *Mathematische Grundlagen der Kartographie*, Bibliographisches Institut, Mannheim, 1969, 2nd ed., 1983.

Jackson, J. E., *Sphere, Spheroid, and Projections for Surveyors*. London, 1987.

Kao, R. C., "Geometric Projections in System Studies," pp. 243–323 of W. Garrison and D. Marble, eds., *Quantitative Geography II*, Studies in Geography 414, Evanston, Northwestern University, 1967.

Keuning, J., "The History of Geographical Map Projections until 1600," *Imago Mundi*, XII: 1–30, 1955.

Lagrange, J. L. de, "Über die Construction geographischer Karten," A. Wangerin, ed., *Ostwald's Klassiker der exakten Wissenschaften*, 55, Engelmann, Leipzig, pp. 1–56, 1894.

Lambert, J. H., (1772): Anmerkungen und Zusaetze zur Entwerfung der Land-und Himmelscharten. *Ostwald's Klassiker der Exakten Wissenschaften*, ed. Vol. 54. W. Englemann, Leipzig.

Löwenhaupt, F., ed., *Johann Heinrich Lambert: Leistung und Leben*, Mülhausen, Braun, 1943.

Lützen, J., Joseph Liouville *1809–1882: Master of Pure and Applied Mathematics*, Springer, New York, 1990.

Maling, D. H., *Coordinate Systems and Map Projections*, London, Pergamnon, 2nd ed., 1992.

Mead, B., *The Construction of Maps and Globes*, T. Horne et al., London, 1717.

Mescheryakov, G., *Teoretischeskie Osnovy Mathematicheskoy Kartografii*, Moscow, 1968.

Mollweide, C, "Ueber Schmidts Projectionsart," *Zach's Monatliche Correspondenz*, pp. 152–163, 1805 (Aug.).

Murdoch, P., "On the Best Form of Geographical Maps," *Philosophical Transactions*, L, II: 553–562, 1758.

Nordenskiöld, A. E., *Facsimile Atlas to the Early History of Cartography*, translation, Stockholm, pp. 84–98, and passim, 1889.

Penrose, R., *The Road to Reality*, Knopf, New York, 2005.

Richardus, P., R. Adler, *Map Projections*. North-Holland, Amsterdam. 1972.

Stevenson, E. L., *Geography of Claudius Ptolemy*, New York Public Library, 1932, (Dover 1992).

Snyder, J. P., "Map Projections—A Working Manual." USGS Professional Paper No. 1395, 1st ed. USGS, Washington, D.C. 1987.

Snyder, J. P., *Flattening the Earth: Two Thousand Years of Map Projections*, Chicago, University of Chicago Press, 1993.

Snyder, J. P., H. Steward, eds: *Biblography of Map Projections*. 1st ed., USGS Bulletin 1856. U.S. Geological Survey, Washington D.C., 1988. Now also online.

Snyder, J. P., P. Voxland, "An Album of Map Projetions," U.S. Geological Survey Prof. Paper, 1453, 1994.

Snyder, J. P., 1994, "How practical are minimum-error map projections," Cartographic Perspectives, 17:3–9.

Thompson, D'Arcy W., *On Growth and Form*, Bonner edition, Cambridge, University Press, pp. 268–325, 1961.

Tissot, A. M., *Mémoire sur la representation des surfaces et les projections de cartes géographiques*, Paris, 1881.

Tobler, W., "Thirty-five years of computer cartograms," Annals, *Association of American Geographers*, 94, 1:58–73, 2004.

Warntz, W., P. Wolff, *Breakthroughs in Geography*, New York, New American Library, 1971.

Wright, E., *Certain Errors in Navigation Detected and Corrected*, London, Moxon, 1657 edition.

Yang, Q., J. Snyder, W. Tobler, *Map Projection Transformation: Principles and Applications*, London, Taylor and Francis, 2000.

INDEX OF MAP PROJECTIONS